Praise for Dr. Gundry and *Gut Check*

"Dr. Steven Gundry has done it yet again—provided his expertise and insight to a crucial health issue, with a fascinating and eminently readable book. His advice should help us all to live healthier, happier, smarter, and longer."

—Dr. Dale Bredesen, author of the *New York Times* bestseller *The End of Alzheimer's*

"There isn't a single aspect of our health that isn't impacted by the state of our gut. In *Gut Check*, Dr. Gundry masterfully guides us through the universe inside of our bellies. You'll discover why your gut health is at the center of your wellness solar system and how to put everything into immaculate order. Read this book. Your body will thank you."

—Shawn Stevenson, bestselling author of *Eat Smarter* and *Sleep Smarter*, and host of *The Model Health Show*

"Dr. Gundry has done it again. In *Gut Check*, he dives deeply into the mysteries of the gut biome and its complex relationship with your overall health. Along the way, you'll begin to understand how important the food choices are in your energy, health and well-being. I highly recommend this book as a strategy in your longevity quest."

—Mark Sisson, founder of Primal Kitchen foods and *New York Times* bestselling author of *The Keto Reset Diet*

"Dr. Gundry does a wonderful job outlining how the different kingdoms of the gut drive our mood, hunger, and even our thoughts. What can be complicated science is explained simply and elegantly with tangible action items for brain health, hormones and gut health. *Gut Check* is an essential book to help you gain insight into your digestive issues and, most important, plan for improvement."

—Dr. Stephanie Estima, DC, author of *The Betty Body*

Praise for Dr. Gundry and *The Plant Paradox*

"It is rare that a book on health and nutrition introduces truly novel concepts, and rarer still that it backs them up with solid scientific data and clinical proof of validity—but *The Plant Paradox* does just that. We all owe Dr. Gundry a debt of gratitude for his insight and dedication to a new level of superb health."

—Dr. Dale Bredesen, author of the *New York Times* bestseller *The End of Alzheimer's*

"*The Plant Paradox* offers a true paradigm shift in the way we think about food. Whether you're paleo, gluten free, low carb, or plant based, Dr. Gundry's profound insights and practical advice will help reverse both autoimmune and neurodegenerative disease."

—Terry Wahls, MD, author of *The Wahls Protocol*

"Once in a generation, a doctor and a book come along that completely change the way we think about food and our health. Dr. Gundry is that physician and *The Plant Paradox* is that book. Following his advice, like I do personally, is life changing."

—Tony Robbins, author of the *New York Times* bestseller *Unshakeable*

"Dr. Gundry is a true trailblazer, always at the forefront of scientific knowledge. *The Plant Paradox* shows the world what 'pioneer thinking' is about and is a must-read book for anyone interested in being as healthy as nature has designed them to be."

—Alejandro Junger, MD, *New York Times* bestselling author of *Clean Gut* and *Clean Eats*

"If you have 'tried everything' but you or a family member are still troubled by stubborn pounds, autoimmune disease, or even heart disease, you owe it to yourself to read this book. *The Plant Paradox* is going to dramatically improve lives—it's that simple."

—Jonathan Carp, MD

Praise for Dr. Gundry and *The Longevity Paradox*

"In *The Longevity Paradox*, Dr. Gundry upends many long-held beliefs about aging, debunking common misconceptions and making the provocative yet persuasive argument that the length and quality of our lives is a function of the health of our microbiome. Practical, informative, and hopeful, this book will change the way you think about aging."

—Mark Hyman, MD, director of the Cleveland Clinic Center for Functional Medicine and #1 *New York Times* bestselling author of *Food: What the Heck Should I Eat?*

"*The Longevity Paradox* offers a fresh perspective on an existential question: how can we live a long life and grow old while still enjoying the physical and mental qualities of being young? What Dr. Gundry shows us is not only that it's possible to thrive as we age, but that the power to do so lies in our own hands (or, more accurately, in our guts). It's an essential read for anybody who's growing older—which is to say, everybody."

—Arianna Huffington, founder and CEO of Thrive Global

"We're often told that future scientific research will provide an elixir that will not only allow us to live longer, but also live healthier as well. But as Dr. Gundry deftly reveals in *The Longevity Paradox*, we have those tools in hand, right now, that can open the door to a longer and healthier life experience. This highly empowering text leverages leading-edge research and provides a user-friendly interface enhancing and extending our precious existence."

—David Perlmutter, MD, author of the *New York Times* bestseller *Grain Brain* and *Brain Maker*

"In *The Longevity Paradox*, Dr. Steven Gundry translates the complex science of aging into a clear, actionable plan. If you want to live longer and healthier, this is the best book on the subject."

—Dr. Dale Bredesen, author of the *New York Times* bestseller *The End of Alzheimer's*

"Dr. Gundry has already helped many of the millions of people who are suffering from inflammatory disorders. Now, in *The Longevity Paradox*, he shows readers how to fend off the inflammation that contributes to disease and decline with age."

—Valter Longo, PhD, director of the Longevity Institute, USC, and bestselling author of *The Longevity Diet*

Praise for Dr. Gundry and *The Energy Paradox*

"Essential reading for the many people who are overworked, burned out, running on empty, and feeling sick and tired all the time. As Dr. Gundry shows, fatigue is not your fate!"

—Arianna Huffington, founder and CEO of Thrive Global

Praise for Dr. Gundry and *Unlocking the Keto Code*

"An epochal piece of research. *Unlocking the Keto Code* is a mind-bending, head-scratching new tome that is going to cause a ruckus . . . in all the right ways. Required reading for anyone on—or even thinking about—a keto diet."

—James Nestor, bestselling author of *Breath*

"Dr. Gundry has done it again. He has the unique ability to take complex scientific information and turn it into simple and practical solutions. His latest book is a must-read for anyone interested in optimal health and longevity."

—Dr. Jack Wolfson, board-certified cardiologist and founder of the Natural Heart Doctor

"Dr. Gundry has always been a skilled disruptor of commonly held beliefs that tend to cause people to struggle with their health. Ketogenic diets, though rooted in long-lived nutrition principles, have led to a startling misconception about how they work. Dr. Gundry, true-to-form, comes sweeping in with *Unlocking the Keto Code* to clarify this nutritional phenomenon like only he can. You'll be healthier, happier, and more empowered by reading this book."

—Shawn Stevenson, bestselling author of *Eat Smarter* and *Sleep Smarter*, and host of *The Model Health Show*

GUT CHECK

ALSO BY DR. STEVEN R. GUNDRY

Dr. Gundry's Diet Evolution

The Plant Paradox

The Plant Paradox Cookbook

The Plant Paradox Quick and Easy

The Longevity Paradox

The Plant Paradox Family Cookbook

The Energy Paradox

Unlocking the Keto Code

GUT CHECK

Unleash the Power of Your Microbiome to Reverse Disease and Transform Your Mental, Physical, and Emotional Health

Steven R. Gundry, MD

With Jodi Lipper

HARPER WAVE

An Imprint of HarperCollins*Publishers*

This book contains advice and information relating to health care. It should be used to supplement rather than replace the advice of your doctor or another trained health professional. If you know or suspect you have a health problem, it is recommended that you seek your physician's advice before embarking on any medical program or treatment. All efforts have been made to assure the accuracy of the information contained in this book as of the date of publication. This publisher and the author disclaim liability for any medical outcomes that may occur as a result of applying the methods suggested in this book.

HarperCollins books may be purchased for educational, business, or sales promotional use. For information, please email the Special Markets Department at SPsales@harpercollins.com.

FIRST EDITION

Designed by Nancy Singer

Library of Congress Cataloging-in-Publication Data

Names: Gundry, Steven R., author. | Lipper, Jodi, author.
Title: Gut check: unleash the power of your microbiome to reverse disease and transform your mental, physical, and emotional health / Steven R. Gundry, MD, with Jodi Lipper.
Identifiers: LCCN 2023039167 (print) | LCCN 2023039168 (ebook) | ISBN 9780062911773 (hardcover) | ISBN 9780062911780 (ebook)
Subjects: LCSH: Gastrointestinal system—Microbiology—Popular works. | Gastrointestinal system—Immunology—Popular works.
Classification: LCC QR171.G29 G86 2024 (print) | LCC QR171.G29 (ebook) | DDC 616.3/3–dc23/eng/20231011
LC record available at https://lccn.loc.gov/2023039167
LC ebook record available at https://lccn.loc.gov/2023039168

23 24 25 26 27 LBC 5 4 3 2 1

To Pearl, Flo and Milt, Bev and Bob:

Thanks for making Penny my soulmate!

CONTENTS

INTRODUCTION

BACTERIAL BRAIN WASHING

What if I told you that free will is an illusion but that instead of a vast, mysterious universe out there controlling our fates, there is actually a vast, mysterious universe within us that we are on the verge of being able to comprehend? By the end of this book, I hope to convince you that this is true, teach you how this universe was designed to guide and support you, and empower you with the ability to restore this universe—which we have unknowingly decimated—to heal anything that currently ails you.

Have you ever seen the movie *Men in Black*? In one scene, there's a little alien named Frank the Pug who tells the main characters that the galaxy they are looking for is here on Earth: "The galaxy is in Orion's belt." The humans are understandably confused. They assume that he is referring to the well-known constellation Orion, which, first of all, is not on Earth, and second, whose belt is made up of three stars. Galaxies have an estimated one hundred million stars. How could three measly stars contain an entire galaxy?

As Frank goes on to explain, "You humans! When are you going to learn that size doesn't matter? Just because something's important, doesn't mean it's not very, very small!"

Over the course of the film, the protagonists come to understand, as I have, that whole galaxies can exist in unexpected places. And that was exactly what the little Frank the Pug was trying to communicate. It turns out that an unfathomably vast galaxy of stars, solar systems, and planets filled with beings is actually hanging off the collar of a cat named Orion. Get it? Orion's belt.

Those characters made the same mistake as humans everywhere. But instead of looking up at the stars, we've looked in all the wrong places for answers about our health and longevity. We've focused our search externally, assuming that the most important things are physically big and therefore must live outside us, whereas in reality the most significant, far-reaching contributors to our well-being and health are actually the smallest. And to find them, we must look inward.

The truth is that inside your digestive system lives a galaxy made up of trillions of bacteria belonging to at least ten thousand different species, plus an as-yet-undetermined number of viruses, fungi, and other microbes. This is your gut biome. You also have an oral biome with seven hundred species of bacteria and a skin biome with a thousand different species. As I've written before, all these living microscopic creatures collectively make up your holobiome. Among them, these microbes contain more than three million genes, whereas the human genome contains a mere twenty-three thousand.

Take a moment to consider just how vast this thing is. There are somewhat more than eight billion human beings on this planet. That means there are *12,500 times* as many bacteria in your own gut as there are humans on Earth. If you prefer plants to humans, consider this: it was recently calculated that there are approximately three trillion trees on Earth.[1] That's more than seven times as many as experts previously believed, despite the billions that humans cut down each year. Even with that discovery, however, there are *ninety-seven trillion* more bacteria in your gut than there are trees on Earth.

When I was in medical school back in the Stone Age, we were taught that the human gut was basically a hollow tube. Food went in, digestion occurred, proteins, sugars, and fats were absorbed, and whatever waste was left over came out as feces. Now we know that our gut is akin to a teeming tropical rain forest with its own diverse ecosystem, communities, and multiple signaling devices, languages that single-celled organisms use to talk among themselves.

Quite startlingly, they also use these languages to tell your mind and body how to think, feel, and behave and how to maintain your skin, muscles, joints, organs, cells, and even the organelles within your cells to keep them healthy—or, conversely, to attack them with inflammation and disease. These billions of single-celled organisms manipulate and control us in unfathomable and downright shocking ways.

It's been only six years since I wrote *The Plant Paradox*. I thought I knew a lot back then, and I'm pleased to say with the passage of time that I was on the right track. But in the time that has elapsed since the publication of that book, scientists have discovered a whole new world of information about the microbiome and the multiple languages its members use to interact with every part of our bodies, to communicate with one another, and, most important, to control the power plants in our cells, our mitochondria. Through these communication signals, they exert control over every aspect of our health, wellness, and longevity. And you are about to learn how to decipher this system and use it to your advantage.

Let me start with a small (literally) example of how these bugs can control us.

Many of us are familiar with the single-celled organism *Toxoplasma gondii*, which is responsible for the disease toxoplasmosis. Pregnant women are told to avoid close proximity to cat litter and to have their partner "scoop the poop" for the duration of their pregnancy because cats can pass along toxoplasma in their feces. If a pregnant woman becomes infected with toxoplasmosis, it can cause severe health problems in her unborn child. But of course, most of us don't stop to consider how this single-celled creature came to live in cat litter to begin with.

Bear with me for a moment. I promise that whether or not you ever plan on getting pregnant or owning a cat, this relates to you in more ways than you can probably imagine.

Toxoplasma has two life cycles. There is a host that the organism ultimately wants to get to, and it uses an intermediary host before arriving at that final destination. In this case, the final objective is a cat. It can be a tiger, a house cat, any kind of cat. Toxoplasma can reproduce only in the gut of a cat—and, like all life-forms, its end goal is to reproduce and pass on its genes to the next generation.

To get into a cat, toxoplasma uses a rodent as its intermediary host. This makes pretty good sense. After all, rodents are famously cats' favorite prey. Just look at *Tom and Jerry*, my favorite cartoon when I was a kid. It seems logical enough for toxoplasma to hang out in the rodent, hoping that it will eventually be eaten by a cat so it can end up in that feline's gut.

But toxoplasma has evolved over millions of years so that it doesn't have to just wait around; it can actually change the rodent's behavior so that it's far more likely to get eaten by a cat. What's that? A single-celled organism can manipulate the actions of a mammal? Yes, it can, and if I teach you anything in this book, hopefully it's that single-celled organisms are far more intelligent than we've ever considered—and that they not only are capable of controlling us but in fact are already doing it. Constantly.

Okay, then, you might be thinking, the toxoplasma must paralyze the rodent or something similar to make it easier feline prey. That would certainly make sense, but it actually chooses a much more complex and nuanced mode of manipulation than that. Unlike brave little Jerry, most rodents are innately afraid of cats. They are repelled by the sight of a cat or even the smell of cat urine. In fact, if you take a rodent that's never been exposed to a cat (or its urine) in its life and expose it to the smell, it will run away. That fear and the associated stress response are hardwired into the rodent for its survival.

Toxoplasma undoes that fear response. It manipulates the fear pathways in the rodent's brain to make it not only less afraid of cat urine but actually attracted to the smell of it. Now, instead of running away, the rodent will approach the cat urine. *Sniff, sniff . . . hey, that smells pretty good!* Uh-oh, here comes the tomcat! And now the rodent is much more likely to end up in its belly.

How does toxoplasma do this? In his lab at Stanford University, Robert Sapolsky, one of my heroes and a professor of biology, neurology, and neurosurgery, studied the brain chemistry of rats that were infected with toxoplasma. What he discovered was shocking.[2]

In both rodents and humans, the amygdala is the part of the brain associated with fear. Toxoplasma infiltrates the nervous system of the rodent, travels to the amygdala, and shrivels the dendrites, which are the branches through which neurons receive information from other neurons. This disconnects the fear circuits in the amygdala. But toxoplasma is even more precise than that; it leaves the circuits associated with other types of fear alone and disconnects only the ones associated with the fear of predators.

Now the rodent is no longer afraid of cat urine and will no longer exhibit a stress response and run away when exposed to its smell. That's pretty good. But how do you get it to be attracted to the urine instead? It turns out that in its genome, toxoplasma has the necessary genes to produce the main enzymes found in dopamine, the neurotransmitter that's all about pleasure, attraction, anticipation, and reward. And it produces that dopamine and sends it to the rodent's brain, activating a different circuit this time: the one that's associated with sexual attraction.

When the rodent smells cat urine now, it no longer experiences a stress response because the pathways associated with the fear of predators have been deactivated. Even better, it really likes the smell because the sexual attraction circuitry is activated instead. The rodent will now run toward danger in the form of cat urine instead of away from it, often leading to its own demise.

This brilliant single-celled organism has completely hijacked the brain chemistry and behavior of a mammal for its own selfish purposes. Pretty darn impressive, isn't it? But these little buggers don't stop with rodents. In 2022, field biologists noted that many of the gray wolves in Yellowstone National Park were infected with toxoplasma and wondered if the wolves' behavior was also being manipulated. It turned out that wolves with toxoplasma were forty-six times as likely to become pack leaders[3,4] as wolves without toxoplasma. Obviously, pack leaders must be bold risk-takers, and toxoplasma infection was causing them to take more risks. But why is toxoplasma bothering with wolves? Because the main predators of gray wolves are cougars, aka mountain lions, aka cats.

By now you might be thinking back to the cat litter situation. What is going on with toxoplasma and humans? When humans become infected with toxoplasmosis, they can get very sick, but many remain asymptomatic while toxoplasma lives in the body. An estimated one-third of the population in developed countries is infected with toxoplasma. Those people are considered asymptomatic because they are not sick, but that doesn't mean that they aren't affected by it.

In fact, it is during this latent stage that toxoplasma begins producing the enzymes that make up dopamine. We may not become attracted to cat urine, but humans with toxoplasma do become a little more impulsive, tend to disregard rules, and are more likely to put themselves into harm's way to save others! So all of those so-called heroes out there may just be acting as tools for toxoplasma.[5] Humans with toxoplasma also have two to three times the likelihood of dying in a car accident due to reckless driving as do people without toxoplasma.[6] That extra dopamine certainly makes us run toward danger.

There is also an interesting link between toxoplasmosis and schizophrenia.[7] We know that patients with schizophrenia have altered dopamine levels in the brain. And if rodents that have toxoplasma are treated with the same drugs that are used to treat schizophrenia, they stop being attracted to cat urine. Pretty wild.

But why would toxoplasma want to mess with a human? After all, it can't reproduce in our gut. In some parts of the world, humans and great apes have always been the main food choice of tigers and other big cats. Indeed, some researchers have looked at our closest relatives, chimpanzees, which are the prey of leopards. Lo and behold, chimps do in fact lose their aversion to leopard urine when they are infected with toxoplasma.[8] And remember, toxoplasma isn't picky about which breed of cats it lands in; it's just as happy reproducing in the gut of a tiger or leopard as it is in Tom's belly.

In other words, toxoplasma uses us (and our close chimpanzee relatives) in the exact same way it does rodents: it eliminates our fear and causes us to run toward danger so that we turn ourselves into easy prey. We think that we're the highest-functioning organism on the planet and that our mind has complete control over our behavior, but in the hands of a simple single-celled organism, we are essentially nothing more than a giant lab rat.

Is this a mic drop moment or what? I assure you, there are plenty more to come.

The thing is, this is not a rare pattern. Like toxoplasma, the other single-celled organisms that use us as their hosts exert their control over us in a multitude of highly complex, sophisticated, intelligent ways.

As Frank the Pug asked, "You humans, when are you going to learn?" The truth is that there's still a lot we don't know about this microscopic galaxy. But we can't afford to wait until we have the full picture to act. For too long, we've overlooked these microbes in favor of the relatively tiny number of cells that constitute what we think is human. Now we are paying the price. There is power in numbers, and those numbers tend to dislike being ignored.

Yet that is exactly what we have done. In fact, we've gone far beyond simply ignoring the majority; we've treated it with outright hostility. Over the past fifty years, we have introduced innovation after innovation that has overlooked, depleted, and destroyed our microbiomes. It's no coincidence that in the same time frame there has simultaneously been a stark increase in major diseases, from obesity to the current mental health crisis to the autoimmune epidemic that I treat in my clinics every day.

In this book, you'll see how each of these diseases, and many more, are directly tied to the ways that modern living has decimated the majority shareholders in our guts. It's safe to say that these microbes are angry, and they're not going to take it anymore. We are all in dire need of a *Gut Check*.

The good news is that the vast majority of the bugs living in our guts do not want us to run toward danger as toxoplasma does. Quite the opposite! They want us to stay healthy and thriving because it is in their own best interest. As far as these gut buddies, as I like to call them, are concerned, we're the cat; they want to reproduce in our guts and pass on their genes.

Our relationship with these bugs is symbiotic. Your body is their home. Ever since the time of Louis Pasteur, we have been taught that these bugs are our enemies and mean us harm, or that at the very least we would be better off without them. But as we now know, we can't live well without them. And if you treat them well, they will in turn take good care of you.

Hippocrates was 100 percent correct more than twenty-four hundred years ago when he said, "All disease begins in the gut." He also believed that a physician's purpose was to be a detective. He proposed that we all have within us (translated from Greek) a "green life-force energy" that provides perfect health. The physician's job, he suggested, is to identify the factors that prevent our green life-force energy from flourishing and teach the patient to remove those causes with no further intervention.

At the risk of sounding a bit woo-woo, I believe that Hippocrates was right about our life force. No matter what is ailing my patients, I investigate until I discover the root cause, which without fail lies in their gut. As I've written elsewhere, once we restore the gut to a state of equilibrium, the disease often abates or even disappears. I don't have to do anything except act as Sherlock Holmes. Our green life force is the galaxy of very, very small beings living behind your belt! And now that scientists are discovering the mechanisms behind these diseases, I propose advancing Hippocrates' theory a step further to say that all diseases can be *cured* in the gut, too. And with the *Gut Check* program, that is exactly what you will be able to do.

So let's get going. We've got a whole galaxy to explore.

CHAPTER 1

YOUR BODY IS A RAIN FOREST

Imagine that you are hosting a big party. As you consider your invite list, there's a lot to think about. You have many different groups of friends from various parts of your life: your old college buddies, your work friends, and some others that you've picked up along the way. Then there are your partner's different groups of friends to consider, too.

It would be nice if everyone got along, but unfortunately, that's not the case. In particular, your work friends and your college friends don't seem to like each other very much. They disagree on some political issues, and previous parties have gotten awkward. A few times, arguments have even broken out.

It might be tempting to hang out with each group separately, but this time, you want to throw a really fun, robust party. Maybe it's a special occasion. You've also noticed that a little competition among groups isn't always a bad thing; it keeps everyone on their toes. Plus, each group brings slightly different qualities to the table, and they tend to complement one another. One work friend makes an amazing artichoke dip, but it's nothing without the (grain-free, of course) crackers that a friend from another group buys at a store near her home. One of your partner's friends has a family member who runs an organic winery, and the wine from there pairs beautifully with a goat cheese from another friend's farm. Any party without all of the guests and everything they contribute would feel incomplete.

So what do you do? You text invitations to all of your guests. Before RSVP-ing, most of them text you back, asking "Is so-and-so coming?" or "Can I bring a plus-one?" or "What can I bring?" Once they feel assured that other members of their group will be there, they agree to attend and confirm what goodies they'll bring.

Believe it or not, something remarkably similar is happening within your gut every minute of every hour of every day. Your gut buddies aren't just getting together to party, though. They work together, play together, and make great efforts to maintain a state of homeostasis—stable equilibrium between interdependent elements—as long as you have a healthy microbiome that includes the right balance of different species that support and challenge one another.

Depending on the mix, they will find a way to work together either to harm or to benefit you, each taking its own unique actions that are all a part of the greater plan. Because of their interrelated functions and the ways they like to compete with one another, diversity is of the utmost importance. But each species will not act until it knows that enough of its members are there to provide backup.

How do they do all this? Well, they talk to one another, of course, using a language that scientists are just beginning to understand.

YOUR HARDWORKING GUT BUDDIES

For just a moment, let's move away from the party analogy and toward another one that I like to use for the microbiome, an ecosystem that is akin to a rain forest in its population density and diversity. In fact, it isn't much of an analogy. The gut microbiome is its own lush, dense ecosystem, teeming with different interrelated, competing, and often mutually dependent species.

Though each of our microbiomes is unique (more on this in a moment), all healthy human microbiomes share the same core patterns, at least from a functional perspective. This means that even if you and I have different species living in our guts, if we are both healthy, our microbiomes will function in very similar ways. Again, this is similar to a rain forest: all rain forests around the world share similar patterns, while each individual rain forest contains its own unique mix of species.[1]

Both rain forests and gut communities contain all kinds of characters: helpful bugs, hardworking bugs, lazy bugs, and a fair share of bad guys, too, whom I like to refer to as "thugs." In the gut, as in a rain forest, these species are interdependent. Each has its own job to do, but they rely on one another in various ways to get their jobs done. No gut bug is an island!

Many bacteria's jobs support one another, and some species cannot complete their job unless another one does its first. Imagine this as an assembly line. The second worker in line (and all the subsequent ones) can't do much of anything until they have the part from the first one. Conveniently, some species of bacteria have similar functional niches and can stand in for one another when needed. So there is more than one worker on the assembly line making the first part. This is how our microbiomes can function similarly, even if their exact makeups are different.

We need this wide mix of characters, even the bad guys. The goal is not to have an entire microbiome filled with happy, helpful, protective bugs. You'll never get rid of the bad guys completely. The idea is to keep them in check with an overwhelming majority of good gut buddies.

You may be wondering: What is the goal? The three most important markers of a healthy microbiome are stability, diversity, and the right balance of cooperation and competition. There is also an interplay among all three of these factors, meaning that the more diverse your microbiome is, the more stable it will become.

Before we get to the three aspects of the microbiome, let's make sure we understand the various roles that gut bugs play. Most of us are aware that our gut bugs help us digest our food. That is true. But scientists are learning more every day about the complex roles they play within this process.

Your gut bugs process your food and deliver vitamins, minerals, and proteins to where they are needed in your body. But it's not as though any old gut bug can process any old type of food. Some bacteria are experts at breaking down starches. Others know how to ferment proteins. Some want to eat only one particular nutrient, such as oxalates[2] (a type of plant compound). Some may need to eat one kind of short-chain fatty acid (SCFA), such as acetate, before they can make another SCFA, such as butyrate.[3] And so on. Each of them has its own niche.

This means that if you don't have the right mix of gut buddies working for you, you won't be able to get all of the nutrients from your food, no matter what kind of diet you follow. As I like to say: *You are not what you eat. You are what your gut buddies digest.* I've known this for a while and discussed it in some of my previous books, but I've learned only recently that we need our gut buddies to process much more than just sugars, starches, proteins, and fats; they also activate some of the most important compounds in our foods, such as polyphenols.

For ages, we've been told that polyphenols are antioxidants that protect our cells from oxidative stress, which is essentially the wear and tear that ages our cells. But I've recently discovered two shocking things about polyphenols: one, they are beneficial for an entirely different reason, which I'll discuss in great detail later, and two, they cannot benefit you at all if you don't have the right mix of gut buddies to process them and make them active. Moreover, polyphenols can regulate which gut buddies are allowed to exist in the gut to begin with and/or prevent them from making harmful compounds that can wreak havoc on the linings of your blood vessels.[4]

But the story doesn't end there. Your gut buddies also control large parts of your hormonal (endocrine) system, your nervous system, and, perhaps most important, your immune system. This means that if you don't have the right microbes, you will be faced with far greater than just GI problems. Your hormone levels will become imbalanced, you will be more likely to struggle with your mental health and suffer from depression and anxiety, and your immune system will get the wrong idea about when and where to launch an attack, leading to widespread inflammation and disease.

Paramount in all of this, part of your microbiome is in charge of "manning the gates" to your body via your gut lining. These "bouncers" decide what is allowed in and what will be kept out. If your gut buddies are working against you or you don't have enough of the right ones to do their jobs effectively, pathogens and other crap are going to end up where they shouldn't—in your bloodstream—leading to a host of problems. (Get it?)

It's no surprise, then, that working closely with my patients to restore their gut biome and gut wall leads to remarkable health transformations, healing everything from skin, bone, and joint conditions to heart disease, Alzheimer's disease and dementia, mental health issues, diabetes, cancer, and all types of autoimmune diseases. In fact, I've come to realize that many of these diseases are actually autoimmune in nature and stem directly from gut dysfunction.

The obvious question, which you're probably already asking yourself, is how exactly your gut buddies exert all of this control. By communicating with every part of your body, of course! Over the past few years, I have focused my research on learning to translate this language, and I am constantly amazed by both its brilliance and its complexity. You will be learning a good deal of this language throughout the book.

First, now that you're beginning to understand the importance of a healthy microbiome, let's take a look at its three defining factors.

THE HALLMARKS OF A HEALTHY MICROBIOME

Ecological Stability

The term *ecological stability* means that a system has the ability to return to a state of equilibrium after a disruption. This is what makes any ecosystem sustainable over time.

Unless it is faced with a major upheaval, the microbiome, like a rain forest, maintains ecological stability. Individuals generally contain their own unique mix of bacterial species in their guts, and that combination remains consistent for long periods of time.[5] In fact, if we were to compare samples from your microbiome taken decades apart, they would be more similar to each other than a sample of yours and mine taken on the same day.

The microbiome's ecological stability is of utmost importance. Patients with stable microbiomes can recover more quickly from a perturbation, whether it's due to a nasty bug such as listeria, a course of broad-spectrum antibiotics, or something else. Though some people's guts respond differently from others, just a short course of antibiotics creates significant disturbances in the microbiome that can last for up to two years![6] The disruption becomes more severe, and it becomes more difficult to return to equilibrium, each time we are exposed to the same antibiotic.[7,8]

What happens if our gut biome cannot return to equilibrium and is therefore unstable? Remember that your gut contains a wide mix of both good and bad bugs. If your microbiome is unstable, the appearance of a few new bad guys could potentially throw off its balance. Those bad guys could reproduce rapidly and find a way to take over. If you're lucky, that will make you feel very sick. Then you know to take action. But if you don't feel the disturbance initially, the bad bugs will have ample time to create an environment that will set the course for a lifetime of chronic disease conditions.

With a stable microbiome, on the other hand, you may still get sick from a nasty bug such as listeria, but the good guys will win the day and return to a state of equilibrium much more quickly. To put it simply, a stable microbiome is far more resilient than an unstable one.[9] It makes sense, then, that a major disruption to the microbiome, creating instability, can lead to disease.[10,11]

Notably, the one time when you do not want a stable microbiome is as an infant. In fact, the infant microbiome is considered to be volatile because it changes so quickly until it stabilizes after about three years of age. Pardon the brief "thread drift" here, but this is important. Contrary to what we believed only a few years ago, during pregnancy, the mother's oral, gut, and vaginal microbiomes affect the health of the fetus in significant ways. The placenta even contains bacteria! These maternal microbiomes all naturally become even more stable than usual during a healthy pregnancy to support the fetus's growth.

During a vaginal birth, the infant interacts with the maternal microbiome even more, and its gut is seeded with these bacteria, which form the foundation of their own microbiomes. Yet after birth, during the first two to three years of life, the infant's microbiome changes rapidly to establish a complex community of gut bugs. This happens through a process similar to ecological succession, the way the mix of species in a habitat changes over time.[12] This involves various communities naturally replacing one another, with each community creating the conditions that allow the next one to thrive.

Back to the forest for a moment: If a large field in the right climate is left alone for many years (which would be unlikely today, unfortunately), it will first become a meadow. The grass will add nutrients to the soil, allowing bushes to grow, followed by trees, which will eventually form a forest. Eventually, this succession will stop, and the system is stable.

The same thing happens in the infant's gut. The early "pioneer" species from the mother undergo a systematic series of turnovers, with each community setting the stage for the next one to thrive. Finally, a complex, stable community is formed. This transformation happens in order to develop and mature the infant's immune system. That's right; during early development and into adulthood, the gut bacteria shape the tissues, cells, and molecular profile of our gastrointestinal and whole-body immune systems.[13]

The infant microbiome also changes to effectively process the infant's changing dietary needs, from breast milk or formula to solid foods. At around the age of two or three, the immune system has been shaped and the toddler begins eating a wider variety of solid foods. The microbiome becomes increasingly rich, diverse, and stable[14] and hopefully stays that way through adulthood.

This long-term stability does not happen by chance. It takes effort on the part of your gut buddies. They work hard to maintain equilibrium and even harder to restore it after it has been disturbed. They do this through feedback loops. Let's say, for example, that one type of bacterium starts to reproduce beyond a certain threshold. This poses a threat to the microbiome's stability, whether the reproducing bacteria are good guys or bad guys.

Luckily, there's a backup plan for that. All bacteria produce metabolites, which are substances created as a result of digestion. These metabolites are used as signaling molecules, and they are more important than anyone previously imagined. In this case, the signal goes out that there are too many of these particular guys around, which triggers changes to the gut environment that make it more difficult for those specific bacteria to grow.[15] This quickly restores balance. Pretty impressive, isn't it?

To take this a step further, many of the bacteria in your gut exist specifically for the purpose of maintaining stability. These "keystone species," as they are called, do not interact with you, their host, at all; they simply work to keep things happy and stable in the gut.[16] You can think of these as being akin to the people you might have stationed outside your big party, checking to make sure that everyone who arrives is in fact on the guest list. Those people don't even attend the party, yet they have a large influence over what happens inside.

The immune system also works to maintain the stability of the microbiome, which is fascinating when you think about it, considering that it's the microbiome that develops the stability of the immune system in the first place. These two systems continue their tight-knit, multidirectional relationship throughout your lifetime, and it is this relationship that determines pretty much every aspect of your health. When the microbiome is disturbed, bacteria signal the immune system to attack, and the immune system works to suppress the bacteria that are overgrowing and causing instability.[17] Once again, equilibrium is restored.

Perhaps most important, diversity, another hallmark of a healthy microbiome, also increases its stability.

Diversity Matters

When it comes to the gut, diversity means that you have a high number of different species that are evenly spread across your microbiome. This is key to a healthy, resilient, stable gut. A diverse gut biome is directly linked to good health and longevity,[18] and low diversity is associated with acute diarrheal disease,[19] inflammatory bowel disease (IBD),[20] liver disease,[21] and cancer.[22]

Research shows that, among other factors, people who are obese have lower gut diversity than lean people do.[23] Low-diversity microbiomes that are associated with disease tend to resemble the volatile microbiomes of infants. In other words, a sick adult has a microbiome similar to that of a healthy baby! This can be explained by another phenomenon seen in the forest—this time in the case of a forest fire. In a forest or the gut, when a disturbance kills off the complex community that creates diversity, the bugs from the early pioneer species discussed previously are more resilient and are able to survive the disturbance.[24,25] This is called secondary succession. It leads the gut to revert to its immature, less diverse state.

Think back to what you read a few minutes ago about the infant microbiome. It starts with pioneer species and then goes through the process of ecological succession before it reaches maturation—all for the purpose of establishing the immune system. This is why infants are more susceptible to disease than adults; their guts have not had a chance to fully develop their immune systems. It makes sense, then, that sick adults have a less diverse microbiome that is associated with an immature immune system. Simply put, your immune system requires a diverse microbiome in order to function properly and keep you healthy.

You also need a wide mix of bugs because each serves a unique and important function, and some of them are real specialists. It's not simply a matter of some gut bugs aiding in digestion while others signal the immune system, for example. Each has its role to play. So the wider your range of gut bugs, the better your gut biome can function as a whole.

Luckily (but not coincidentally), your gut has developed a backup plan for this. Though many gut bugs are specialty workers, some species can fill in for one another when needed. They have what's called functional redundancy; they know how to do the same job. For instance, various species can process complex carbohydrates and produce the necessary metabolites. If one of the species is killed off by antibiotics, another species will start reproducing rapidly to take over the wiped-out gut bug's job.[26] This helps keep things moving along, but it still leads the gut to become less resilient due to its reduced diversity. After all, if something threatens the backup species, there's no other species around to fill in for it.

If this happens, you will develop what's called low-diversity dysbiosis, an imbalance of gut bacteria that is linked to disease. Notably, the specific gut bugs that tend to be missing in cases of low-diversity dysbiosis (Lachnospiraceae and Ruminococcaceae) are responsible for fermenting complex sugars into short-chain fatty acids (SCFAs), such as butyrate.

Butyrate is an incredibly important SCFA that is linked to immune system function,[27] but the roles that butyrate plays in the body have been widely misunderstood and are more far-reaching than we ever imagined. I will discuss this in great detail later on. For now, it's important to know that low-diversity dysbiosis leads to a decrease in butyrate and other important SCFAs.

As I said, a healthy, diverse gut biome includes a few bad guys and even some microbes that have long been considered parasites, including protists and helminths (better known as worms).[28] Many of them aren't just innocent bystanders, either. They have their own roles to play, too. For instance, some protists and helminths stimulate a beneficial immune response in people who suffer from allergies.[29] Some brave souls have even started intentionally ingesting these parasites to calm their immune systems, but I don't necessarily recommend that you try this at home.

It seems that a healthy, diverse, stable gut biome needs to have some bad guys present in order to keep the good guys on their toes. This leads me to the final hallmark of a healthy biome, and it's one that I admit surprised me: a healthy sense of competition between microbes that is balanced by mutually beneficial cooperation.

Cooperation and Competition

Let's return to the party analogy for a moment. Remember the friend who brought the delicious dip and the other who contributed the crackers? These two items are basically useless without each other. Nobody wants to shovel dip into their mouths with their fingers, and the crackers are bland and dry without the dip. These individual items are mutually beneficial; each benefits from the presence of the other.

This is an example of cooperation, and your gut buddies do the same thing. In their case, cooperation means that various species become healthier and function better in one another's presence. They need one another to survive and thrive. This is not a "predator/prey" or parasitic relationship in which one species benefits while the other suffers—although some do function this way. As in all healthy relationships, these ones are mutually beneficial and truly symbiotic.

It's taken your gut microbiome millions of years to coevolve in order to cooperate this way. The mechanisms behind this cooperation are fascinating and incredibly complex, often involving many different species at once. In fact, when two species are studied together, it's actually unlikely that they will cooperate in a way that's mutually beneficial; it is more likely when three or four or more species are studied together. Again, this is akin to a rain forest and is yet another reason that diversity is so important.[30]

This was observed when three species of bacteria were grouped together in a lab. Each species performed a job that was required for any of the three to survive. And it seemed that that was no coincidence. Each produced more of the necessary compounds than needed, creating a surplus for their cooperative partners.[31]

One of the primary and simplest ways that our gut buddies cooperate is through cooperative digestion. This is when the compounds that one type of bacteria produces through its digestion process become the food for another. As the bacteria digest the compound, they produce another one that becomes the food for a third species. And so on.

Another interesting piece of this puzzle is the fact that sometimes bacteria actively secrete these compounds, often along with other waste products, whereas other times they hold on to them. In the latter case, these compounds are released only when the bacteria die. This is one reason that dead bacteria are so important. Yes, you read that correctly! You need plenty of living *and dead* bacteria in your gut. Talk about diversity. Dead bacteria carry nutritive compounds for living bacteria, as well as essential information. More on this later.

But perhaps the most fascinating way that bacteria cooperate with one another is through "quorum sensing." Hold on to your hat, because this is where it gets really interesting. We all know what a quorum is, right? It's the minimum number of group members that need to be present at a meeting to make any decisions that are made valid. Well, a gathering of gut buddies has a quorum, too, and it doesn't act until enough members of the group are present.

Your gut buddies can't really "see" how many others are present. After all, they don't have eyes. So they've created their own ingenious way of taking attendance: they use signaling molecules to communicate with one another. This is part of the language that I referenced earlier. It's much like your friends texting you to see who else is coming to the party.

Your gut buddies use these chemical signals to synchronize their activities in a process that involves producing, releasing, detecting, and responding to those compounds. The last step—responding—is threshold dependent, meaning that the bacteria do not respond to the signals until they detect a certain level that tells them they have reached a quorum.[32]

Even more impressive, bacteria don't use this quorum-sensing process unless they are embarking on an action that can be done effectively only as a group. In this way, quorum sensing allows single-celled bacteria to act like much more advanced, multicellular organisms when necessary.[33] Smart little buggers!

But make no mistake; not all gut buddies cooperate with one another. And it's actually important that they don't. Let's think back to the party analogy for a moment. Maybe you have friends from two different groups who don't actively dislike each other, but they seem to compete for your favor and your friendship. This isn't necessarily a bad thing. Having a bit of competition motivates each of them to be better. When one friend cracks a joke, the other tries to say something even funnier. When one friend shares an interesting piece of trivia, the other tries to top it with an even more fascinating fact. You get the idea. A healthy sense of competition can make people better. And it can make your gut buddies better, too.

To put it simply, in a healthy microbiome, your gut buddies need some collaborators and some competitors in order to function at their best. One reason for this is that if every species cooperated with the others, they would become interdependent. Then, if one species decreased in number for any reason, the other species would die alongside them, destabilizing the whole system.[34]

Lo and behold, there is a built-in solution for this, too. When certain competitive species are paired, they function more efficiently than they do separately.[35] This is just like your competitive friends. Or it may be more helpful to think of two competitive athletes who each forces the other to work harder and get stronger. This decreases the likelihood of the strengthened species dying out and taking its cooperative species down with it. They really do have a solution for everything!

• • •

If only we humans were smart enough to leave environments like the rain forest or your inner ecosystem alone to thrive in a naturally beautiful, peaceful state of homeostasis. But of course, we have decimated our guts, just as we've killed off so much of the rain forests, throwing off this delicate balance. Have no fear, however. Armed with the knowledge of what your gut buddies need in order to thrive, we are going to reseed, feed, and nurture your gut back to being a stable, diverse ecosystem with the right mix of cooperation and competition to create equilibrium.

First let's take a closer look at your gut buddies' language—and this time, not the one they use to communicate with one another, but the one they use to profoundly impact the energy centers of your cells.

CHAPTER 2

IT TAKES TWO

By now you're beginning to understand how your gut buddies live, work, and communicate with one another and to appreciate how important they are to your health and well-being. I expect that you're at least a little bit impressed by the design of your inner ecosystem and that you're becoming motivated to work with your gut buddies to help reestablish your gut's beautiful, natural state of homeostasis. The truth is that so far we've uncovered only the tip of the iceberg.

See, your gut buddies don't just stick to themselves in their own happy ecosystem. In fact, they work and communicate with every other system in your body. The most important and closest knit of these relationships is the one that they share with their sisters, which live and work within your own cells.

A TALE OF TWO SISTERS

If you're new to my books, welcome! I'm incredibly happy that you're here. If you've read my other books, some of this information will serve as a refresher. Either way, it will be handy to have it at the forefront of your mind before we move on.

Most of you who took high school biology class are aware of the "power plants" in our cells, the mitochondria. Mitochondria have a fascinating history. The prevailing theory of their origin is that they evolved from engulfed bacteria. If you rewind back to two billion years ago, the world was full of different types of bacteria—but some other fledgling types of cells were also in the mix. As the theory goes, one of those cells, likely a precursor of the eukaryotic cells that make up most of life on Earth, engulfed those bacteria. They started to work together, forming a symbiotic relationship in which both organisms benefited. The bacterium helped the cell to respire, or use oxygen to make energy. The cell, in return, gave the bacterium a home, helping to keep it safe from the elements. Over millions of years, those bacteria evolved into mitochondria.

Despite the fact that mitochondria exist inside our cells, they've never fully given up their bacterial roots; they are actually quite similar to the gut bacteria that inhabit your microbiome. In fact, just like your gut buddies, mitochondria have their own DNA. They can divide at the same time that their host cells divide, but they can also divide to make more mitochondria at any time they please via a process called mitogenesis. As you will soon learn, their ability to replicate themselves without the rest of the cell's having to divide is critical for you, your health, and your fate.

Today, your microbiome and mitochondria are still tethered together by their shared bacterial past. They stay in touch via signaling molecules called postbiotics. These are generally produced by the microbes in your gut, but there are also many of them that you can swallow. Your gut microbes vigilantly monitor everything that takes place in your body. They are in a great position to do so, as they regularly receive information about the state of affairs from both your immune system and your nervous system. They then pass along messages to the mitochondria about how much energy they need to produce via postbiotic signaling molecules.[1] The information mitochondria receive from your microbiome influences how much energy they produce.

This is one reason why traditional keto, low-carb, or high-protein diets, which restrict the intake of plant-based fiber, can lead to side effects such as fatigue and brain fog. That fiber is essential to producing a thriving microbiome that in turn produces these essential postbiotics. We'll discuss this in more detail soon, but first, let's take a closer look at how our energy factories work.

HOW MITOCHONDRIA MAKE ENERGY

The technical name for the conversion of food and oxygen into energy is *cellular respiration*. It's a process that occurs over and over again in every single mitochondrion in the body—and, if you recall, your body houses trillions of them. You can think of cellular respiration as a bit like an internal assembly line. It takes several steps to convert glucose (or proteins and fats, for that matter) into adenosine triphosphate, or ATP, which is our cellular energy currency.

As any *Star Trek* fan can and will tell you, human beings are carbon-based life-forms. We also consume carbon. All of the food we eat, whether it's made up of sugars, amino acids, or fats, eventually breaks down into a bunch of carbon molecules. Those carbon molecules make their way inside the cells, where they are scooped up by the mitochondria to kick off the energy production process. This is the start of the Krebs cycle (sometimes called the citric acid cycle), the series of reactions that turns those carbon molecules into ATP.

Once the carbon molecules pass into the mitochondria, they begin a fascinating dance with protons and electrons, some of which come from water. You may recall that these are charged (think electrified) particles. The protons carry a positive (+) charge, and the electrons carry a negative (–) one. Those protons and electrons are then ushered across the inner mitochondrial membrane to the innermost part of the mitochondria. There they go through a series of chemical reactions known as the electron transport chain. It's complex stuff—but, simply explained, that chain helps ramp up the charges of those particles. As their charge increases, the excited electrons and protons become like proverbial hot potatoes, becoming more and more electrified as they jump from one level of charge to the next.

Things are so "hot" in this situation that you can think of this process a bit like a bunch of twentysomethings heading out for an evening at the latest, greatest nightspot. Think of a single mitochondrion as the hottest new club in town. If you read my book *Unlocking the Keto Code: The Revolutionary New Science of Keto That Offers More Benefits Without Deprivation*, you will recall that I referred to this as the Mito Club. For those of you who haven't read it or need a refresher, this concept is so important that I'll reprise it here so that we are all on the same page.

THE MITO CLUB

This hip new spot has one main entrance that lets you in and a one-way revolving door at the rear where you can exit. (There are also some emergency exits. The Mito Club does want to stay in code, after all, but we'll talk more about those later. At this point, there's only one way in or out for the Mito Club's patrons.)

The Mito Club is crowded, hot, and packed to the gills with hundreds of protons, electrons, and other molecules including oxygen and hydrogen. Because the Mito Club can get so full—anyone who is anyone is trying to get in—a doorman oversees the entrance. It's his job to admit only so many people. But even with the doorman working hard, patrons can barely get to the bar without bumping into at least a dozen other people. And, as in a real club, there are plenty of protons and electrons that are there with the hope of connecting (coupling) with an oxygen molecule.

Some do manage to couple with that desirable oxygen. They link arms, head to the back revolving door—and make a lot of ATP upon exiting. It's a bit like the way water creates the power to turn a millstone when it goes over a mill wheel. When the positively charged protons, now coupled with oxygen, pass through the back door turnstile of the mitochondria's membrane together, they make energy. Furthermore, in the process of exiting, the protons leave behind carbon dioxide (CO_2). In this scenario, you can think of the CO_2 as the beer bottles and other trash the proton patrons toss before heading out the door with their dates.

It's a typical, jam-packed Saturday night when, all of a sudden, a lot of the electrons decide to leave. That leaves a bunch of protons, which had hoped to connect with some oxygen, milling about and realizing that the odds of connecting with oxygen are low. The frustrated protons see the Exit sign in the distance and make a beeline for the door. There, some of the protons may meet up with some wayward oxygen molecules and finally manage to couple up. Then, as the new couples push through the turnstile, they produce more ATP. But most of the other protons won't be so lucky. They are going to exit the club alone and unfulfilled. They won't be making any ATP tonight.

I should mention that this process isn't quite as simple as I just made it sound. In mitochondria, the multistep process of trying to combine oxygen with protons to make ATP can lead to more than just the production of some CO_2. Unfortunately, it's not just the protons that want to couple with oxygen; the electrons are all jazzed up, too! When electrons instead of protons end up coupling with oxygen, they produce reactive oxygen species (ROSs), which include the free radicals that we health experts talk about so much.

Think of ROSs as being a bit like the exhaust from your car's engine. In the Mito Club analogy, I think of them as the patrons who have overindulged and start throwing punches. The bouncers will eventually deal with them, but that doesn't mean that they don't cause a ruckus! ROSs are behind oxidative stress, which damages mitochondria and, as a consequence, cells. You've probably heard of ROSs and oxidative stress before. Both have been implicated in aging and chronic disease.

Now, some ROSs are okay. The Mito Club wouldn't be the place to be if there wasn't a little excitement. In small amounts, they act as signaling molecules, sending messages to help keep your cells healthy. It's only when they are produced in excess that they become a problem. So when too many electrons and oxygen molecules couple up, it can damage mitochondria. Even worse, if the Mito Club's bouncers don't find a way to keep the ROSs in check, they can induce apoptosis—literally, the cell's explosive and immediate death. As you can imagine, too many fights, too much drama, and the club would have to be shut down.

The Mito Club's two main bouncers are melatonin (yes, the sleep hormone you've heard so much about) and glutathione, a well-known antioxidant. They help keep the ROSs in their sweet spot, just enough of them milling about to perform their signaling duties but not so many as to harm the cell. As you can imagine, the Mito Club likes to have plenty of those bouncers to make sure things don't get out of hand. But to prevent things from getting out of hand, there has to be another fail-safe system in place. This is where mitochondrial uncoupling comes into play.

UNCOUPLING FOR HEALTH

Although today, many people use the term *uncoupling* to describe leaving a romantic relationship, mitochondria have their own way of divorcing the burning of fuel (metabolism) from the production of energy (ATP). This is called mitochondrial uncoupling.

Let me explain how uncoupling works using the Mito Club analogy. As you know, this juke joint is *the* place to be, and the line to get inside has gotten long and longer. Things are getting hot inside, and, at a certain point, the protons are no longer interested in coupling up with the oxygen molecules. They just want to get out and either try a new place or call it a night. The Mito Club has only one exit in the back, but with the big crowd, that exit is easily bottlenecked. Then someone pushes open one of the emergency exits, and bam! The protons rush out the door, so invigorated by their new freedom that they head down the block to try their luck at coupling somewhere else.

Now that some room has opened up in the Mito Club, the tension dissipates and the patrons can enjoy themselves again. In fact, the protons and oxygen molecules now have room to start coupling up again. Furthermore, the doorman can now admit some of those that have been waiting outside. But within just a few minutes, the club is jam packed again. What should the owner do? He's got unhappy customers, both those who are packed tight inside and all those outside who are waiting to enter. He needs a new game plan.

With so many patrons milling about outside, it's clear that the cell could use another club, or more mitochondria, to keep up with the demand. This process is called mitogenesis, or the addition of more mitochondria to the cell. Under certain circumstances, a cell will make more mitochondria to handle the workload. (As you may recall, mitochondria have their own DNA and can divide when they need to, regardless of what the rest of the cell may be doing.) Most experts will tell you that there are only two ways to make more mitochondria: by fasting and/or exercising. But I'm here to tell you that there are actually several other ways to invoke mitogenesis—and you can get the resources you need to do so by following the eating tricks in this book.

So the club owner decides to build more clubs. He's perfected a recipe for success with the Mito Club. But to create the new venues, he needs a bank loan. Where can he go to get it? As it turns out, he can tap into the body's fat stores to provide the resources to build the new clubs (mitochondria) and start producing all that feel-good energy again.

Okay, you might be thinking, *but why would the fat stores give out their currency? Something isn't adding up.* You'd be right to think that. It takes the involvement of unique proteins, nudged by ketones or other "uncoupling compounds," to open the side doors, promote mitogenesis, and tell the fat stores to open up.

In 1978, the physiologists David G. Nicholls, Vibeke S. M. Bernson, and Gillian M. Heaton, researchers at the Buck Institute for Research on Aging, discovered that mitochondria have built-in "emergency exits" for the various players participating in the electron transport chain. Those exits are controlled by uncoupling proteins.

Today, we know that there are five total uncoupling proteins, named UCP1 to UCP5. They all reside within the inner mitochondrial membrane and allow protons to exit under certain circumstances. Like the Mito Club patrons sneaking out the side door, our mitochondria can allow uncoupled protons to leave the cells' energy-producing power plants—and waste calories in the process!

As it turns out, ketones, as well as other molecules that I will discuss, send messages to the mitochondria to open those emergency exits, or uncouple, and make less ATP. In the process, the mitochondria perform a caloric bypass, wasting calories instead of using them for fuel, and induce mitogenesis, making more of themselves.

By uncoupling, your mitochondrion each ends up creating less ATP than it otherwise would. This may sound like a negative thing, but it actually has tremendous benefits. As first described by Martin D. Brand, PhD, a prolific researcher also at the Buck Institute for Research on Aging who studies the mechanisms of energy transformation in the human body, mitochondrial uncoupling is all about protecting the mitochondria. Remember, making ATP damages the mitochondria. So uncoupling protects the health of the mitochondria from those nasty ROSs.

Having healthier, better-protected mitochondria is key to every aspect of your health. Its importance cannot be overstated. I would go so far as to say that at the end of the day, you are only as healthy as your mitochondria.

After uncoupling and mitogenesis, two healthy mitochondria can make more ATP together than one damaged mitochondrion can on its own. Think of a dogsled: If only one dog is pulling the thing, you won't be able to go very fast. Plus, the dog will tire out before too long, so you won't get very far, either. For every additional dog that you hitch to the sled, you'll go farther and faster because the dogs are now sharing the burden. The only downside is that now you have to feed more dogs. That is going to require more food.

The same is true of mitochondria, but in this case, feeding more mitochondria is actually another benefit since uncoupling allows you to waste calories and take fat out of storage to feed your new mitochondria. Further, uncoupling mitochondria produces heat through a process called thermogenesis that promotes weight loss, vitality, and good health.[2]

Bottom line: You want your mitochondria to uncouple and make more of themselves. How can you make that happen? This is where the mitochondria's sisters come into play.

UNCOUPLING COMPOUNDS

Let's return to the inner mitochondrial membrane, where the uncoupling proteins reside. They are basically the guards at the side entrance of the Mito Club that can open the emergency exit doors, triggering the process of uncoupling. But they don't just open up willy-nilly or even every time the mitochondrion gets taxed or overloaded. They wait for a signal from their sisters.

These signals come in the form of uncoupling compounds, which include polyphenols, short-chain fatty acids, and ketones. This does not, however, mean that you can just eat a bunch of polyphenols, drink some apple cider vinegar (which contains SCFAs), and consume MCT oil (which generates ketone production in the liver) to reap the full benefits of mitochondrial uncoupling. If only it were that simple! Look back at the name of this chapter: "It Takes Two." Regardless of the compound that you are using to uncouple your mitochondria, your gut buddies are, and always will be, a huge part of the uncoupling equation.

Polyphenols

As I said earlier, polyphenols are not antioxidants, as we've been led to believe. Like antioxidants, they do protect our cells, but, unlike antioxidants, they do it by uncoupling mitochondria. That's if—and only if—your gut buddies process them for you first. For years, my fellow researchers struggled to understand how polyphenols function within us, since they are so poorly absorbed. Now we know that your gut buddies are the middlemen (or women!), and this is the missing link that eluded us for so long.

It turns out that polyphenols are one of the primary (and favorite) food sources of many of your gut buddies. They love the darn things, and as they digest them, they transform them, making them into more absorbable and bioactive compounds.[3] It is in this new form that they can act as mitochondria uncouplers.

One way to think about the difference between antioxidants and polyphenols is that antioxidants repair cellular damage from ROSs after it has already happened. This is like the cleaning crew that comes in after the Mito Club is closed for the night. But after they've been processed by your gut buddies, the polyphenols do something even better: they trigger mitochondrial uncoupling to prevent the damage from happening in the first place.[4]

You might be asking yourself why polyphenols even exist. As you know, humans need oxygen to make energy, but that oxygen can damage mitochondria by creating ROSs. This is our original catch-22: can't live without oxygen, can't live with it! Similarly, plants need the photons in sunlight to make energy (ATP), but those photons can damage the plant's mitochondria, which are called chloroplasts. Plants produce polyphenols to uncouple and protect their own mitochondria.

Believe it or not, every fall you get to enjoy seeing all those polyphenols in the glorious color display of fall foliage. The gorgeous yellows, reds, oranges, blues, and purples in the leaves were there all along, doing their job of uncoupling the leaves' mitochondria. You just didn't see them until the dark green chlorophyl disappeared. And dark-colored fruits and vegetables? Their colors are from polyphenols, too. If we consume them, they can protect our mitochondria the same way they do the plants'—but, again, only if our gut buddies are there to help us absorb them.

After they've been processed by your gut buddies, polyphenols help us out in other ways, too. For starters, they send a signal to sirtuin 1 (SIRT1), an enzyme in the cell nucleus that repairs DNA and protects it from damage. SIRT1 is deeply involved in metabolism, brain function, and aging, and when polyphenols stimulate the overexpression of SIRT1, it broadly enhances health and life span.[5] So your gut buddies are essentially empowering polyphenols to tell the cells to protect both the mitochondria and themselves—a double whammy.

Resveratrol (RSV), one type of polyphenol that has gotten a lot of attention for its presence in wine, is one of many polyphenols that can activate SIRT1. Indeed, as my wife likes to remind me, moderate champagne consumption is good for you! It has been shown to decrease the risk of cardiovascular disease.[6] But RSV has an even bigger impact on the brain and metabolism than it does on the heart. In rats with neuropathy similar to Alzheimer's disease, RSV is able to inhibit memory loss by creating neural malleability in the hippocampus. It also improves metabolic disease and increases life span.[7]

Independent of SIRT1, RSV and other polyphenols also activate AMP-activated protein kinase (AMPK), an enzyme that activates glucose and fat uptake by mitochondria so they can produce more energy. This leads to neurogenesis (the birth of new neurons) and mitochondrial biogenesis (the birth of new mitochondria).[8] So these powerful polyphenols help you create new, flexible neurons with new mitochondria to power them, all while preventing them from being damaged by ROSs. No wonder your gut buddies think that polyphenols are the best things since sliced bread. (Sorry, bad comparison.)

RSV is so powerful that many researchers are looking at how it can be used to treat or prevent neurodegenerative and metabolic diseases. Ironically, they often conclude that the main obstacle to successful treatment is its low bioavailability,[9] meaning that we cannot absorb it well enough to reap all the potential benefits. Well, of course not. We need our gut buddies to make it absorbable.

Another important type of polyphenol is called ellagitannins. If you have the right mix of gut buddies, they can use ellagitannins, which are found primarily in berries, pomegranates, and walnuts, to produce a metabolite called urolithin A. When urolithin A is given to worms, it increases their life span by 50 percent! Interestingly, this is the exact same effect on life span as is created by periodic fasting or calorie restriction, which I'll discuss in detail later. Urolithin A increases life span via an uncoupling mechanism that triggers mitophagy, the recycling of old or damaged mitochondria, as well as mitogenesis, the creation of new mitochondria.[10] In humans, the mitochondrial effects of urolithin A have been shown to beneficially impact muscle health and reduce cartilage degeneration and pain in people with osteoarthritis.[11]

That all sounds pretty good, but before you rush out to buy berries, pomegranates, and walnuts, there's something you should know. In studies of superold centenarians, about 50 percent of them have the right combination of bacteria to manufacture urolithin A. If it sounds as though they found the fountain of youth in their gut, you are on the right track of where this book is heading. Spoiler alert! Only about 20 percent of the general population has the right mix of bacteria in their gut to produce urolithin A from polyphenols. This means that you can eat all of the pomegranates in the world (yum!) yet get none of this important metabolite into your system.[12]

I'll say it one more time: it takes two. You and your gut buddies need to work together for the sake of your health and longevity. If you are lacking one or more "keystone species," the ecosystem begins to fall apart, and you will fail to reap the benefits of healthy eating.

Once again, however, polyphenols are not antioxidants. In fact, there are only two antioxidants in the mitochondria: melatonin and glutathione. And wouldn't you know it, plants make those, too, to repair any damage caused by photons during their own energy production process. Of course, this means that many of the same plants contain both polyphenols and antioxidants—a one-two punch of mitochondrial protection. Hey, maybe, just maybe, a plant-based diet is good for us. But as with the example of urolithin A above, if you don't have the right mix of gut buddies, all of your "healthy" eating might be for naught. Who would have thought? But don't fret. This book is all about getting your gut into check.

One thing that continues to fascinate me is how our ancestors seemed to have such incredible natural wisdom about what was good for them. This is something that today we seem to sorely lack. Think about the spice trade, for example. As I've written before, spices contain both polyphenols and antioxidants. Wars have literally been fought over spices, and I refuse to believe that it was just because people wanted spicier food. Throughout history, people have been willing to die for really only one thing: drugs. (Okay, and maybe sex, too.) In their own way, spices and other plants are incredibly powerful medicinal drugs. I only wonder how those people knew.

Short-Chain Fatty Acids (SCFAs)

Short-chain fatty acids (SCFAs) are another main category of uncoupling compounds.[13] They include acetate, butyrate, and propionate. Unlike polyphenols, which your gut buddies process for you, SCFAs were once believed to be made directly in the gut when the gut buddies fermented dietary (prebiotic) soluble fiber. But not so fast. Remember, it takes two. In this case, it may take even more than two.

In their research at Stanford University, Erica and Justin Sonnenburg showed that humans fed a high-fiber diet did not see an improvement in their gut microbiome biodiversity or their levels of inflammation. *But* when they supplemented with fermented foods such as kefir, kombucha, yogurt, kraut, and even vinegar that contained postbiotics and ate the prebiotic fiber, their gut microbiome diversity increased and their inflammation markers decreased.[14]

This is a point that I'll keep returning to throughout the book. We've all heard a million times how important it is to eat or supplement with plenty of dietary fiber. But like polyphenols, dietary fiber isn't all that good for you in and of itself. The more salient point is that it's really good for your gut buddies—but if and only if they have the signals and the precursors to make use of it. In order for you to benefit from polyphenols or dietary fiber, you need your gut buddies to digest them for you. But they can't do it alone. There is another step that has long been overlooked.

It turns out that your gut buddies need fiber to be prefermented before they can process it for you. Through fermentation, yeasts and bacteria produce intermediary SCFAs, including formate, succinate, and lactate. Unlike SCFAs, intermediary SCFAs do not play any direct role in your body. They merely serve as a food source for other bacteria, particularly butyrate-producing bacteria.[15] This means that your butyrate-producing bacteria can't make butyrate for you until after its friends make an intermediary SCFA for it to eat.[16] Spoiler alert: fermented foods are lousy sources of probiotics (living bacteria and yeast), but they are loaded with intermediate SCFAs. This means that when it comes to benefiting from dietary fiber, fermented foods are the missing link.

Of course, indigenous cultures around the world have been fermenting foods since long before anyone saw bacteria and yeasts under a microscope. The first documented case of human-induced fermentation was in North Africa around 10,000 BCE, when fermented milk was used to create yogurt. Three thousand years later, the Chinese began brewing a beverage made with fermented fruit. Jump ahead another three thousand years, and the Egyptians were using yeast to ferment dough and make leavened bread. At around the same time, Chinese people began pickling vegetables and using moldy soybean curds as an early form of antibiotics. And so on for centuries. Guided by their intuitive sense of what felt right in their bodies and the collected wisdom of their societies, these populations were able to leverage the powers of fermentation and polyphenols together, significantly increasing their bioavailability and antioxidant properties[17] long before Louis Pasteur "officially" discovered fermentation in 1856.

But dietary fiber isn't the only food source that your gut buddies can ferment to produce SCFAs. Certain specialized gut buddies can also make SCFAs, including the most important one, butyrate, using the mucus produced by the epithelial cells lining your gut.

As I will discuss in great detail, your gut lining is protected by a layer of mucus. It is there to trap potential invaders from the outside and keep them from getting into your bloodstream. But that mucus is also a valuable nutrition source for an essential type of gut buddy, who can use it to produce SCFAs. This species is called *Akkermansia muciniphila*, which literally means "mucus loving."[18] These gut buddies eat the mucus layer, which stimulates further mucus production, ultimately strengthening and protecting the gut lining. If you've read my other books, you already know a bit about these guys. They're some of the most important gut buddies out there (or should I say in there), and you'll learn even more about them later on.

Though *Akkermansia* can eat mucus to produce butyrate, the butyrate also tells the epithelial cells to increase their mucus production. This leads to improved intestinal barrier function.[19,20] Plus, when *Akkermansia* are well fed and taken care of, they keep the gut layer nice and strong, with impermeable, tight junctions holding the wall together. They also reproduce, so that there are more of them to eat the mucus, make more butyrate, and signal the epithelial cells to make more mucus.

There is no oxygen way down deep in your colon. What's a poor colon cell to do? Butyrate to the rescue! Butyrate is the sole source of nourishment for colon cells. If there is no butyrate, there will be no happy colon cells, no barrier to prevent unwanted particles and bacteria from getting loose—creating the ideal conditions for cancer cells to grow. In just this one example—and there are plenty more to come—Hippocrates was right: all disease begins in the gut.

Remember the main reason that I'm talking about butyrate to begin with. It's a signaling molecule that tells your mitochondria to uncouple, thereby protecting the health of the mitochondria and the cells themselves. When you have strong, healthy, plentiful mitochondria in the cells lining the gut, your gut lining will also be strong and healthy! This is precisely how butyrate improves and even repairs intestinal barrier function.[21]

SCFAs also aid in weight loss, in part thanks to the ways that uncoupling "wastes" calories for you. When rats are fed a high-fat diet, SCFA supplementation helps them avoid gaining weight, compared to rats on the same diet who are not given SCFAs.[22] Acetate, another SCFA, also helps regulate the appetite by triggering the gut to release hormones that signal to the brain whether you are hungry or full.[23,24,25]

Butyrate in particular is not just a powerful mitochondrial uncoupler; it also has a widespread impact on your entire immune system. It tells the epithelial cells when to release pro-inflammatory cytokines and when not to,[26] which reduces chronic inflammation in the gut. It triggers the release of the antioxidant glutathione to repair cellular damage from ROSs.[27] And it even modulates the function of dendritic cells, a special type of immune cell.[28,29] Butyrate also increases the number of regulatory T-cells (immune cells) in the intestinal lining and increases secretory IgA (SIgA) production.[30] SIgA is the most prominent antibody produced in the body to protect against infection. Germ-free mice without the gut buddies to produce butyrate have a tenfold reduction in IgA.[31]

SIgA is unlike other antibodies, which kill off specific invasive bacteria. Instead, it works in a dynamic process, taking into account the delicate balance of your internal ecosystem. It can alter the behavior of a specific strain of bacteria, for example, by preventing its pro-inflammatory behavior, or it can kill off that strain completely, depending on a host of other factors. SIgA is unique in that it affects the balance of the whole microbiome.[32] It is an important way of maintaining homeostasis.

Basically, SCFAs are a messaging system, a language that your gut buddies create and send to their sisters to let them know what's going on down in the gut. If things are a bit dodgy or invaders are coming, this signal tells the immune system to attack. If things are going well, it tells the immune system to stand down and relax. It tells the mitochondria when to uncouple, get healthier, and start making more energy, and then it sends the message for glutathione to go in and clean up any mess that was made in the process.

Recall for a moment how SCFAs are produced in the gut: through the process of fermentation. Fermentation is the breakdown of a substance by bacteria or yeasts, and hopefully you are beginning to get an idea of how important the fermentation process in your gut really is.

Scientists have known for a long time that we need bacteria to break our food down for us, but they are only now learning about all the beneficial metabolites that are made during the fermentation process and how they communicate with the rest of your body.

Ketones

The one main category of mitochondria uncouplers that is not made or processed in the gut is ketones, but that does not mean that your gut buddies are left out of this equation entirely. Ketones are organic compounds that are made in the liver from free fatty acids liberated from fat cells and can also be made in the liver from ingested medium-chain triglycerides (MCTs), a type of dietary fat. For years, we thought that ketones were an important secondary form of fuel. Of course, many people still believe this. The simple version of this theory is that mitochondria typically use glucose to make ATP, but when there isn't any of that around, they burn fat, in the form of ketones, instead.

This myth is behind all of the hype about the keto diet. But there are a few problems with it. First of all, in order for your body to produce enough ketones to make an impact, it can't have access to other forms of fuel. This means that you either have to fast for at least twelve hours a day or eat only foods that are "keto friendly" and therefore lacking the carbohydrates that are present in most plant foods.

More important, however, it turns out that ketones aren't actually a good source of fuel. Instead, they are signaling molecules that tell your mitochondria to uncouple. This is the real reason that people lose weight on the keto diet! The ketones tell the mitochondria to waste fuel in order to protect themselves from damage and to make more of themselves to pick up the slack. So although ketones are uncoupling compounds, they're not the easiest or most beneficial ones to access.

There are some food sources that help generate ketones, but not many. People often make the mistake of thinking that coconut oil will give them plenty of ketones. In reality, there are different types of MCTs in coconut oil, each containing anywhere from six to twelve carbon atoms. (C6 has six carbon atoms, C8 has eight, and so on.) Simply put, the fewer the carbon molecules, the greater the ketogenic effect. So in order to get adequate ketones, you want more C6 and C8 than C10 and C12. Since C6 smells like goats, C8 is your best and most palatable option.

Coconut oil is made of 55 percent MCTs, which sounds pretty good if you stop there. But if you keep looking, you'll find that it is 42 percent C12, which has no ketogenic effect! This means that only 13 percent of coconut oil is made of the three more beneficial types of MCTs combined.[33]

Alternatively, you can consume MCT oils, which generally have a higher percentage of the more beneficial MCTs. These oils do have a place in the *Gut Check* program. But remember that ketones are not a great fuel source, and there are other powerful compounds that not only uncouple mitochondria but also nourish the gut and offer other benefits.

In addition, you need a healthy gut and the microbes those bacteria produce in order for your body to make ketones in the first place. That's right; your gut buddies are involved here, too. Butyrate is a building block of a type of ketone called beta-hydroxybutyrate (BHB). And you already know that you need your gut buddies to make butyrate. So even when ketones are the signals telling your mitochondria to uncouple, the message is still essentially coming from their sisters in your gut.[34]

Micro-RNA and the Gut

Micro-RNAs (miRNAs) are single strands of RNA molecules that are synthesized in the nuclei of your cells and function in the cytoplasm. Your cells use miRNAs to control gene expression. They have a direct impact on important cellular processes, including apoptosis (cell death), proliferation (cell growth and division), and cell differentiation. Deregulation of miRNA is implicated in many types of cancer,[35] rheumatoid arthritis,[36] and autoimmune disease.[37]

Until recently, scientists believed that miRNAs could work only on human gene expression. But they have recently learned two additional and fascinating things about miRNA. One, intestinal epithelial cells produce miRNAs and send them to the mitochondria to regulate mitochondrial gene expression.[38] Two, miRNAs that are made by bacteria in our microbiome also exist in extracellular vesicles that are absorbed and circulate throughout the body. These extracellular miRNAs can enter bacteria and regulate their gene expression, just as they do to yours! This plays a large role in shaping and modulating the makeup of your microbiome.[39]

It appears that miRNAs are another signaling system. When mice are lacking a specific miRNA-processing enzyme called endoribonuclease Dicer, their gut buddies start to run rampant, and the composition of their gut biome falls completely out of balance. They also have a reduction in epithelial tight-junction molecules that are normally regulated by gut buddies. This leads to cellular infiltration of the colon and susceptibility to colitis.[40] When they undergo a fecal miRNA transplant, gut homeostasis is restored and their colitis improves.[41]

The relationship between your gut buddies and miRNAs is bidirectional, meaning that it goes both ways. miRNAs affect the gut biome composition, and gut buddies also affect miRNAs. This is yet another reason it's so important to have a balanced ecosystem within your gut. Defects in miRNA processing can lead to the growth of cancer cells and dysregulate the microbiome to create an environment that is more favorable for those cancer cells to grow, reproduce, and thrive in.[42]

How does this happen? When your gut is out of balance and certain bacteria overgrow, they can degrade miRNAs, leading to tumor growth and increased inflammation, among other diseases. On the other hand, if all is peaceful and balanced down in the gut, your gut buddies will produce butyrate, which, among its other talents, can alter miRNAs to suppress the growth of cancer cells.[43]

Further, germ-free mice have a higher abundance of miRNAs

than mice that are colonized with bacteria. And when the colonized mice are given antibiotics to wipe out their gut buddies, their level of miRNA goes up significantly.[44]

You may be wondering, *Isn't increased miRNA a good thing?* The short answer is: sometimes. There is often a Goldilocks effect at play when it comes to this stuff. You don't want too much or too little miRNA. Instead, you want a balanced ecosystem with all of the elements working together in harmony. Mouse studies don't show whether increased or decreased miRNA is good or bad, but they do prove the direct relationship between the microbiome and miRNAs. This is yet another layer of the seemingly endless onion that together we are going to unpeel.

GASOTRANSMITTERS

Your gut buddies also produce something called gasotransmitters during the fermentation process. These are important signaling molecules from your microbiome. Gasotransmitters are basically transmitters (chemical messengers) made of gas. You (or the people you live with) may have a negative association with the gasses that are made in your gut, but they actually have beneficial effects on your entire system. Gasotransmitters include nitric oxide (NO), carbon monoxide (CO), and hydrogen sulfide (H_2S), to name three. Yes, you have these gasses, which are often considered toxic, in your body. Further, you absolutely want them there.

Believe it or not, gasotransmitters are another type of signaling system between your gut buddies and your mitochondria. Although each works slightly differently in your body, they all regulate mitochondrial function through uncoupling and by stimulating mitogenesis, the birth of new mitochondria.[45] They form yet another layer of the complex communication system among the bacterial sisters residing inside you.

Hydrogen Sulfide (H_2S)

Scientists used to believe that H_2S was toxic until they discovered its function as a signaling molecule in the body. Hydrogen sulfide (with that famous rotten egg smell) is created by the gut's fermentation of sulfur-containing compounds. It works on neurons and plays an important role in nociception, your nervous system's process of understanding noxious stimuli (heat, cold, mechanical force, and chemical stimulation).

When you experience pain, your gut buddies produce H_2S and send it to your brain to let them know that you're hurt. The H_2S then activates nociception neurons in the brain, which leads to the release of inflammatory cytokines and growth factors to heal the damage.[46] It makes sense, then, that disrupted nociception signals are associated with significant alterations in the microbiome. And when nociceptors are removed, the result is a defective tissue-protective reparative process.[47] Neurons do not get the signal that you're in pain and need to heal. (By the way, nociception provides the basis for "gut feelings" or "gut instincts." Our gut buddies produce H_2S, enabling us to interpret pain and discomfort.)

H_2S is also linked to memory. It accelerates activity in the hippocampus, the center of emotion, memory, and the autonomic nervous system in the brain. It also increases synaptic plasticity,[48] strengthening the junctions among neurons that enable them to communicate. In other words, your gut buddies send H_2S as a signal to the cells in your brain, telling them "Hey, guys, talk to each other!" This enhanced communication between neurons improves memory.

It comes as no big surprise that there is a direct link between H_2S and mental health. Having too little H_2S in the hippocampus causes a condition called chronic unpredictable mild stress (CUMS). This condition is associated with a loss of dendritic spines in the hippocampus and leads to depressive behaviors. This is fascinating, since your neurons need those dendritic spines to—well, talk to one another. H_2S works by sending the signal for the cells to communicate, but without enough H_2S present, they can't. It's not all that surprising, then, that increasing H_2S levels in the brain corrects depressive symptoms associated with CUMS. Amazingly, it does so within just a few hours![49]

In addition to its effects on the brain, H_2S signaling helps prevent heart disease through its role in vasodilation and the process of angiogenesis, the birth of new blood vessels.[50] It also reduces inflammation in the gut and promotes tissue repair in the gastrointestinal (GI) tract, along with protecting the mucus layer lining the gut.[51]

However, as with most things, there is a Goldilocks effect with H_2S and other gasotransmitters. You don't want either too much or too little of it. H_2S was originally considered to be a toxin because too much of it inhibits mitochondrial function.[52] But now we know that you don't want too little of it, either.

One key to maintaining ideal levels of H_2S is to reduce your intake of animal protein. Restricting the amino acids derived from animal protein results in balanced H_2S production.[53] On the other hand, a diet high in animal protein throws this balance in the wrong direction, damaging your precious mucus layer and disturbing your microbiome diversity.[54] This is one reason that following a typical keto diet may negatively impact your longevity plans.

Nitric Oxide (NO)

Your gut buddies produce nitric oxide (NO) through the fermentation of ammonia. NO helps regulate dopamine, which as you know is an important feel-good chemical that is released in the brain.[55] Our brains release dopamine to motivate us to do something, to reward us for doing it, and to help us learn from it. So when we are faced with a task, our gut buddies recognize that we need a little extra motivation and produce NO for us that then travels to the brain and triggers the release of dopamine. Pretty nifty.

In addition, NO plays a role in developing the nervous system. It triggers the growth of nerve fibers and new neurons and the formation of new synaptic connections.[56] Our gut buddies use this communication tool to encourage us to complete tasks and to form the "hardware" in our brains so that we can learn from them.

Just recently, a study out of Case Western Reserve University School of Medicine, University Hospitals Cleveland Medical Center, and Harvard Medical School looked at the effects of NO secreted by gut bacteria inside tiny worms.[57] They found that the NO attached to thousands of proteins, completely altering the worm's ability to regulate its own gene expression. When the worms had an overabundance of NO-producing bacteria, it silenced genes that are critical for development. The worms developed malformed reproductive organs and died. Though this is not a practical concern for us—NO-producing bacteria will not naturally overgrow to that extent in our gut—it illustrates the tremendous power of a balanced (or imbalanced) microbiome.

In humans, NO is directly linked to mental health. Too much NO damages synaptic connections and neurons[58] and is present in patients with severe bipolar disorder (BP). Wouldn't you know it, lithium, the first line of treatment for BP, regulates NO levels.[59]

In addition to its impact on dopamine, NO mediates the effects of various neurotransmitters that make you feel good, including norepinephrine, serotonin, and glutamate. Patients with depression have altered NO levels in various brain regions, cerebrospinal fluid (CSF), blood, and exhaled gas.[60] Like lithium, many antipsychotic medicines work in part by altering levels of NO in the brain.[61] It's ironic that so many effective medications work by altering the gut biome. Why not go straight to the source and heal the gut itself? Well, that's what we're going to do.

NO's most impressive effects happen at the blood vessel level. It's actually so important that the endothelium of the blood vessels themselves can produce NO. It is responsible for dilating blood vessels, and a lack of it is a major driver of high blood pressure and kidney damage.[62]

But wait, this is a book about your microbiome. The gene that codes for the enzyme that converts dietary nitrate to nitrite to then make NO is called endothelial NOS, or eNOS. (I promise, this will not be on the test.) eNOS often becomes dysfunctional as we age, but there is a microbial work-around: a robust oral microbiome can convert dietary nitrates into NO.[63] Unfortunately, several human studies show that most commercial mouthwashes destroy your oral microbiome. You can just hear the ads now: "Kills 99 percent of odor-causing bacteria in your mouth."

Maybe we should have more truth in advertising: "Your fresh, minty breath comes at the cost of your needing to take a high blood pressure medicine!"

Periodic tongue scraping, however, promotes a more diverse oral microbiome. This can in turn improve your NO production. Once again, there is scientific research confirming the benefits of this ancient, seemingly archaic practice—and the fact that you are the end product of your microbiome.

As a precursor of something I'll talk more about later, fructose, the main type of sugar in fruit, reduces NO synthesis, and high fructose intake is linked to heart fibrosis, cardiac hypertrophy, and vasoconstriction.[64] Just because fructose is "natural," it doesn't mean we should be eating it year-round in large quantities! Reducing NO synthesis is just one reason this sugar is bad news.

Carbon Monoxide (CO)

In the brain, carbon monoxide (CO) protects cells from harmful agents in a process called cyto-protection, which leads to a reduction in cell death. It also reduces neuroinflammation by regulating cellular redox responses.[65] This is the balance between creating ROSs and removing them with antioxidants. You already know that this protects the health of mitochondria.

Interestingly, NO signaling is associated with the anti-inflammatory effects of CO in the microglia,[66] which are the immune cells of the brain. At this point, it should come as no surprise that your gut buddies use these various communication systems to work together in the most intricate ways.

CO also plays an important role in fetal brain development. In rats, low prenatal exposure to CO leads to a variety of neurological disorders, hampers the functions of neurotransmitters including dopamine, and affects sexual behavior. These changes are linked to deficits in motor tests, learning, and memory.[67] This is one reason it's so important for expecting mothers to make sure they have the right gut buddies in place, working together and doing all of their various jobs! Of course, we all need to do that (and are going to), too.

• • •

You just read a lot of information about the various molecules your gut buddies produce to communicate with and control the rest of your body, particularly your mitochondria. Yet, as I keep saying, there is still so much more to learn. We will continue to work on translating our gut buddies' language as we go. But first, we need to address perhaps the most important result of mitochondrial uncoupling: maintaining a healthy, impenetrable gut wall.

CHAPTER 3

I HAVE A CRYSTAL BALL—AND IT'S YOUR GUT WALL

Would you assume that I was exaggerating if I told you that when they first arrive at my office, nearly every one of my patients is suffering from leaky gut, also known as intestinal permeability? This condition results from a breach in the gut wall that allows harmful compounds to leak into your bloodstream and your body. Leaky gut is now an epidemic, affecting more people than we've ever before imagined. It is both a result of a damaged internal ecosystem and a contributor to nearly every major disease.

Hyperbole? Well, what if I told you instead that almost all of my patients are suffering from diseases due to widespread inflammation in their bodies? Would that sound more acceptable? This is also true, but, unbeknown to you, most health care professionals, and internet health pundits, "widespread inflammation" is a manifestation and consequence of having a leaky gut, so it's just another name for the same issue.

Unfortunately, this means that you can eat all of the "anti-inflammatory" foods in the world, but it won't really help much with inflammation. For those of us in California, eating anti-inflammatory foods to quiet inflammation is like fighting a forest fire with a garden hose! At best, maybe you can tamp down some small amount of inflammation this way, but unless you address the root cause, you'll never be able to keep up. The inflammation will just keep coming due to the root cause: leaky gut.

To return to the "crystal ball," when I measure how a patient's gut wall is functioning, which I can do with blood tests that measure leaky gut, such as anti-zonulin IgG, anti-actin IgG, and anti-LPS antibody tests, I can literally see their future. If their gut wall is intact and their microbiome is homeostatic, they are likely to live a long, healthy, disease-free life. If the gut wall is permeable, however, they will almost certainly suffer from inflammation, pain, and disease.

The good news is that it is entirely possible to restore the gut wall to all of its former glory, not only reducing inflammation but also reversing disease in the process. The turnarounds that I see regularly in my patients when they follow the *Gut Check* program are nothing short of remarkable, and I assure you that the same thing can happen for you.

THERE'S A BREAK IN THE WALL

Let's start with a brief refresher on how your gut and the gut wall work. Your gut begins at your mouth and nose and continues all the way down to your anus. I like for you to visualize this tube as your skin turned inside out. Think about it: your skin is there to protect you from the outside world, but it also is capable of absorbing a lot of the things that you smear on it, including skin care products and sunscreens (more on those later).

Now think about the last time you got a cut or a splinter. Remember all the redness and swelling that occurred for a while? That's inflammation produced by your immune system guardians (such as white blood cells) attacking foreign invaders such as bacteria and proteins that gained access to your body through the cut or the hole the splinter made. Those foreign enemies got past your skin's protective barrier, and it not only shows but also hurts.

Now let's compare your gut wall and its functions to your skin. Your gut lining, like your skin, should protect you from the foreign things you swallow or inhale, including bacteria, viruses, molds, and worms (yes, worms), as well as from foreign proteins that are part of a plant's defense system against being eaten. (More on this shortly, too.) But your gut wall has to absorb a whole lot more than your skin does. In fact, it's responsible for absorbing all of the nutrients, vitamins, minerals, and other things that you need to function—all the while keeping out everything you don't want and/or that wants to harm you.

It's a tall job, indeed, especially considering that although the lining of your gut has the surface area of a tennis court, that lining is only one cell thick. Yikes! If you think about it, this sounds like a pretty bad engineering design, but it's actually much more complex.

Specialized white blood cells that make up a full 60 to 80 percent of your entire immune system are positioned along that gut lining like troops manning the walls of your fortress. They are there so that when something slips through the lining that shouldn't, they will launch into action and attack the foreign invader, just like what happened in your skin with that cut or splinter.

Inside your gut, your microbiome (with the help of stomach acid and enzymes) breaks down everything you swallow into amino acids, fatty acids, or sugar molecules, which then approach the gut lining, hoping to exit the gut and enter your bloodstream. First, they have to get through several barriers or checkpoints. The first is a thick coating of mucus overlying the gut wall. Next, at the gut wall, they undergo a whole-body scan, much as you do when you go through security at an airport.

If they are deemed safe, your gut lining cells bite off one molecule from each of the digested foods and pass it through, releasing it into a huge vein that carries nearly everything directly into your liver. If things check out at your liver, off they go into your bloodstream. Everything else remains in your gut, where it should be, and is expelled as waste or used to feed the other half of you, your microbiome. Actually, as you already learned, it's much more than half!

But let's back up a second. Why is the gut wall lined with that coating of mucus to begin with? As you are probably well aware, mucus is sticky. In fact, it is composed of mucopolysaccharides, a fancy name for "lots of sugar molecules." If an invader that does not belong in your body makes it all the way to the mucus layer in its efforts to get out of the gut and into your body, the mucus's job is to trap it so it can't get through.

Some of the foreign molecules are proteins that bind with specific types of sugar molecules, many of which are in your mucus. There they remain, bound up in the mucus.

However, if the wrong type of molecules or bacteria make it past your mucus layer, through the gut wall, and into your body or bloodstream, the immune system responds. Remember, a full 60 to 80 percent of your immune system is stationed right there at your gut wall, armed and ready to attack.

The next question is: How does your immune system know which molecules belong and which do not? Those immune cells contain scanning devices called toll-like receptors (TLRs), or, as I call them, "tiny little radars," which mediate the immune response. They are like bar code scanners. They recognize the structure of specific molecules and scan the ones coming through the gut lining to make sure they are safe. When a TLR recognizes an invader, the immune cell releases inflammatory hormones called cytokines. These cytokines signal to the rest of the immune system that there's trouble down in the gut and they should be prepared to launch an attack, too, if the troublemakers show up in their vicinity.

In a healthy body, this system works brilliantly. But when your gut is permeable, invaders regularly end up in places they shouldn't be. Your immune system attacks these invaders over and over again. And all of the cytokines produced lead to widespread inflammation, the root cause of disease that you now see is really a symptom of leaky gut.

What causes the gut lining to become leaky in the first place, triggering all of that inflammation? One culprit is lectins, which you're already familiar with if you've read my other books. Lectins are a type of protein found in many plants that evolved as a defense mechanism to keep the plant or its seeds from being eaten by predators. In response, we humans evolved to have our own defense systems: our microbiome, which loves eating lectins, and mucus, not just along our gut lining but also in our nose, mouth, and esophagus. Now, lectins are some of those sticky, sugar-seeking proteins I mentioned earlier. They are attracted to the mucus, which binds and traps them. Another great system.

The problem is that many of us don't have enough mucus lining our gut walls. Why not? Because we don't have the right mix of gut buddies living in our ecosystems. Remember, you need the gut buddy *Akkermansia* to eat that mucus and produce butyrate, which in turn stimulates the gut lining cells to produce more mucus.

When we are lacking homeostasis in our internal ecosystem, as so many of us unfortunately are, we may not have enough *Akkermansia* and other butyrate-producing bacteria to maintain our gut lining and effectively protect us from invaders. Nor do we have the right mix of lectin-eating bacteria to form another defensive line against the onslaught. So the mucus that we do have gets used up from constantly having to bind and trap so many lectins and other outsiders that are trying to break through. The mucus that is meant to be a renewable resource begins to run dry, leaving us vulnerable to invaders.

It gets even worse. Once they get past the now-depleted mucus layer, lectins can bind with receptors along the gut lining and produce a compound called zonulin, which breaks the tight junctions that hold your gut lining together. What used to be a sticky, impenetrable border is now a lot less sticky because of a lack of mucus, plus it's full of holes that allow any invader in that wants to get through!

The invaders can be lectins, pathogenic bacteria, or lipopolysaccharides (LPSs), which are fragments of cell walls from dead bacteria. As I've mentioned before in previous books, I typically don't swear, but I can't resist calling LPSs "little pieces of shit" because that's literally what they are. Finally, if the wall of the gut has gaps, undigested food particles that normally would never make it across the gut wall without first being broken down and absorbed appear to our immune system to be foreign invaders and are attacked, as well.

Imagine one of my patients being startled to learn that as a result of all this, her immune system had made antibodies to broccoli, which she ate a lot of, thinking it was one of the world's healthiest foods. Her leaky gut had led to undigested broccoli being attacked by her immune system over and over. When your immune system faces the same enemy repeatedly, it creates proteins called antibodies to protect you from it. As part of the immune response, antibodies attaching to their target cause inflammation. Imagine an anti-inflammatory food such as broccoli actually causing inflammation! (Don't worry, the story has a happy ending. After following the *Gut Check* program, both she and her immune system love broccoli once again.)

DEAD MEN MAY TELL NO TALES, BUT DEAD BACTERIA DO

As I was doing the research for this book, I came across some fascinating information about dead bacteria. Yes, you read that right. Dead bacteria actually contain important information for your gut buddies, playing an important role in quorum sensing and other bacterial group efforts. Intriguingly, it appears that although dead and living bacteria both send critical messages to their friends in your gut, the information they share is slightly different.

This is a fascinating part of your gut buddies' language and yet another reason that fermented foods play such a critical role in your health. Most fermented foods contain dead bacteria as well as the postbiotic "messages" that the bacteria produced during the fermentation process (back when they were alive, of course). Thus, a recent human study showed that a diet high in fermented foods dramatically increases microbiota diversity while decreasing inflammatory markers.[1]

Another recent study looked at the effects of supplementation with live versus dead (pasteurized) *Akkermansia* on mice.[2] It showed that both types of supplements were beneficial to the microbiome, but in different ways. Both strengthened the gut lining, but the dead bacteria were actually more effective in that regard than the living ones were! The stronger gut lining led to a greater reduction in inflammation in the mice that received the dead bacteria. The living bacteria, however, had stronger effects on the modulation of gene expression related to fatty acid synthesis, energy homeostasis, and the immune response. The living bacteria also had a greater impact on helping to create homeostasis in the gut.

Clearly, the natural cycle of bacterial life and death is a critical component of how these tiny beings exert their control over us. It's almost as if they had accounted for the fact that they would inevitably die and worked it into their communication system. Are you impressed yet?

Just as your living gut buddies respond to signals from dead bacteria, the TLRs on your immune cells respond to dead bacteria that cross the gut wall—this time in a not-so-good way. The TLRs recognize dead pieces of bacteria (LPSs) as invaders and respond by launching your immune system into action.

As a quick aside, it is interesting to consider the fact that fermenting foods containing lectins is one of the best ways to reduce their lectin content so that you can safely eat them without damaging your gut wall. During the fermentation process, bacteria eat the vast majority of those nasty lectins! Is it possible that they do this intentionally to protect their buddies in your gut? At this point, I wouldn't put it past them.

There are many ways to protect and restore your gut wall, which we will explore shortly. But the most important and effective way is to make sure that the mitochondria in the cells lining your gut keep themselves healthy through uncoupling and mitogenesis. In every part of your body, the healthier your mitochondria are, the healthier your cells will be. And the healthier your cells are, the healthier your tissues and organs will be. This, of course, applies to your gut wall, as well.

On the other hand, when the mitochondria inside your cells become damaged to the point of no return, the cells die and explode in a process called apoptosis. This throws cellular debris, including pieces of the cell walls and mitochondrial walls, into your bloodstream. Remind me again what mitochondrial walls are? Bacteria! And as far as your TLRs are concerned, they might as well be LPSs. TLRs call your immune system into action, and this leads to—you guessed it—inflammation.

How can we avoid this? After all, cells die all the time. That's true, but there is another way for cells to die, this time in a process called autophagy, which literally means self-eating. In this process, the cellular pieces, including the mitochondrial components, are recycled into new cells with new mitochondria, so they are not exposed to the immune system. Autophagy helps prevent apoptosis from occurring[3] and is a critical component of homeostasis.

You can think of autophagy as cells dying from old age and apoptosis as cells dying from disease. In the latter case, it causes inflammation, contributing to your own aging and disease! Once again, how can we help our cells die of natural causes after a happy, healthy, long life? We can make sure that the mitochondria inside them are healthy and protected, primarily through uncoupling. This is true of every cell in your body, and it is of primary importance in your gut wall.

TALES FROM THE CRYPTS

It's common knowledge that our gut wall is the size of the surface area of a tennis court, but most of us don't stop to wonder why. It's because of the microvilli, fingerlike membranes containing your intestinal wall cells. Because these cells are so tightly packed together, the microvilli along your gut wall look like a shag carpet. I think of them as being like roots that are embedded in soil. And just like the soil in the ground, your "soil" is meant to be packed with bacteria and fungi that foster the absorption of nutrients and minerals.

At the base of each microvillus are crypts that hold a pocket of gut buddies and undifferentiated stem cells that are there to replicate and replace dead cells in the gut wall, keeping it strong and healthy. Since the lining is only one cell thick, replacing those cells when they die is of the utmost importance. Scientists believe that these crypts evolved to protect the precious stem cells.[4]

There is a specific balance of gut buddies that are meant to live in the pockets of those crypts along with the stem cells. They are there to enhance nutrient absorption. Using the same metaphor, they are the soil. Interestingly, patients with colon cancer have an altered mix of gut buddies in their crypts.[5] This is just one of the many links between cancer and the gut that I'll discuss in much greater detail later.

Because the stem cells in the crypts are so important, there is also have a type of cell called Paneth cells in the crypts that help to protect and defend them. These cells contain TLRs to identify pathogenic bacteria, and they secrete antimicrobial peptides (AMPs) to kill them off once they are spotted. This protects the stem cells and helps maintain gut homeostasis.

The enterocytes that line your gut also produce AMPs. This enhances gut wall function, which should come as no surprise based on what we know about how thin the gut wall is that they are protecting. Indeed, mice that are missing a specific type of AMP called cathelicidin-WA have a mucus layer that is too thin, making it easier for pathogens to get through.[6]

Your gut buddies have a bidirectional relationship with AMPs. The AMPs help shape the microbiome by killing off bacteria that are pathogenic or overgrowing, but if you do not have a healthy microbiome, you will not produce adequate numbers of AMPs. This is because the metabolites produced by your gut buddies—including, you guessed it, butyrate—are signals that tell your cells to produce AMPs.[7] This explains why treatment with broad-spectrum antibiotics, which we'll talk much more about later, reduces AMP production, allowing pathogenic bacteria to take up residence in the gut.[8] Another shocker (not really): patients with irritable bowel syndrome (IBS) have reduced AMP production.[9]

Okay, so if (and that's a big if) we have the right gut buddies in place, the system can protect the stem cells in the crypts. Then, when a stem cell down in the crypts is called into action, it has to proliferate, differentiate into epithelial cells, and move out of the crypt and up the roots to replace the damaged or dead cells.[10] What calls that stem cell into action? You guessed it: the gut buddies that are hanging out with them down in the crypts. Their metabolites, including good old butyrate, signal to the stem cells that their services are needed.[11] They then spring into action to proliferate, differentiate, and travel up out of the crypt and across the microvilli to protect the gut wall.

Let me tell you a quick story that may help you think about this in a different way. Back in the 1970s, I was a fellow at the National Institutes of Health (NIH) in heart surgery. We performed a heart valve replacement on a young man from China. The operation was tough, and we faced a lot of problems. The patient survived, but after the surgery, he didn't wake up. He remained in a coma for weeks. All of his vitals were stable, but he wouldn't wake up, and we didn't know why.

After a few weeks, the patient's brother flew in from China. He came into the ICU, bent down, and whispered something that we couldn't hear into his brother's ear. Almost immediately, our patient woke up.

We couldn't believe it. Had he sensed his brother's presence, or was there something else going on?

Later, the patient's brother explained that they came from an ancient line of kung fu practitioners. My understanding is that in their practice, they learn to put themselves into a trance to protect themselves in the face of imminent danger. Only a fellow practitioner knows the secret "code" that tells them that they are safe now and it's okay to wake up. The man had simply whispered that code to his brother as the "all clear" sign. Not only did the patient wake up, but he fully remembered the events of the two weeks when he had been "comatose." He even greeted the nurses who had been caring for him by name and correctly related incidents that had occurred in the ICU during those two weeks.

This may seem a little "out there," and my colleagues at the NIH certainly weren't interested in writing up a case report in a medical journal about it. Even if you don't believe the story, allow it to illustrate my point about the crypts. The stem cells are down there in hiding, and your gut buddies send butyrate and other metabolites as codes that tell them, "Come on out. We need you, and it's safe."

The problem, of course, is that without the right gut buddies present, there is no one there to send the signal. And just as we couldn't wake up our patient without his brother being present, your stem cells will not be able to differentiate and restore the gut lining unless they receive the right signals. The gut buddies have to be there to send the code. Otherwise, there could be tons of damage to the gut wall, but the very stem cells that are waiting in the crypts to fix it would never know!

Here's a clear-cut example: In one study conducted on mice, crypts that were devoid of bacteria had abnormal stem cell activation and an increase in apoptosis. When pathogenic bacteria were allowed to populate the crypts, the mice developed lethal sepsis. And when the crypts were repopulated with the original mix of gut buddies, homeostasis returned, and the crypts regained their regenerative abilities.[12]

Another piece of this puzzle is vitamin D, which we know also has a bidirectional relationship with the gut biome.[13] The level of vitamin D in your body affects your mix of gut buddies, and you need the right mix of gut buddies to synthesize vitamin D for you. This is why, in one study, mice without a vitamin D receptor developed severe diarrhea, rectal bleeding, and severe weight loss and died.[14] The researchers conducting the study found a disruption in the epithelial "tight junctions" in those mice and concluded that vitamin D plays a critical role in mucosal barrier homeostasis by preserving the integrity of those tight junctions.

Now we are beginning to understand why. It turns out that without sufficient vitamin D, the stem cells in our crypts fail to get the signal to differentiate into epithelial cells. This is likely because vitamin D increases the abundance of *Akkermansia*, along with two other major butyrate producers, *Faecalibacterium* and *Coprococcus*.[15] So if you don't have enough vitamin D, you won't have enough of the right gut buddies to send a signal to the stem cells in the crypts. To complicate this even further, because of the bidirectional relationship between vitamin D and the gut, if you don't have the right gut buddies, you won't be able to synthesize vitamin D![16]

Once again, what was once a beautifully complex, multilayered system has devolved the further away we've gotten from our own "roots." Though we once got plenty of vitamin D from the sun, we now spend most of our time indoors or slathered in sunscreen. Whereas we once ate tons of fermented foods, polyphenols, and antioxidants, our diets are now void of uncoupling compounds. And that doesn't even begin to touch on the fact that our gut biomes are now a desert wasteland. Of course, leaky gut is now an epidemic! We are lacking everything that we need to maintain, restore, and regenerate our gut lining. But together, I promise that we will get it back.

BRINGING IN THE REINFORCEMENTS

Because your gut lining is of utmost importance, there are many systems in place to protect and restore its function. In order to rejuvenate our gut walls to their former glory, we must leverage all of these mechanisms. They include the following.

Intestinal Alkaline Phosphatase (IAP)

Another cornerstone of gut barrier health is intestinal alkaline phosphatase (IAP), an enzyme that is secreted by our enterocytes. It is present throughout the GI tract and is especially prevalent along the microvilli. One reason IAP is so important is that it is an additional line of defense against LPSs, those harmful dead pieces of bacteria that send your immune system into overdrive when they cross the gut barrier.

IAP protects you from LPSs in two ways. First, it keeps them from crossing the gut wall and circulating through your body, and second, it breaks down their chemical structure, or "bar code," by removing one of their two phosphate groups. This reduces their toxicity one hundred-fold. The broken-down, or dephosphorylated, LPSs still bind to TLRs, but they do not lead to the same immune system reaction, thereby reducing the inflammatory response.[17]

But the impact of IAP on gut barrier function doesn't end there. It also promotes gut barrier function by upregulating the proteins that protect the tight junctions along the gut wall[18] and has a direct influence on the microbiome itself. It promotes the growth of helpful gut buddies by preserving homeostasis and even helps restore the state of homeostasis after the gut has been wiped out with antibiotics[19] or disrupted by a traditional Western diet.[20]

Interestingly, IAP levels naturally increase during early childhood, when the microbiome is rapidly becoming more diverse. And it appears that IAP plays an important role in modulating the gut from the very beginning.[21] IAP levels tend to decline with age, and this is generally accompanied by increases in gut permeability and systemwide inflammation. But when mice were given supplements of IAP, their gut permeability was improved and they had a significant reduction in their levels of inflammatory cytokines.[22]

Okay, so we know that we want this stuff. How can we get our bodies to make more of it? Your IAP level is dependent on the foods you eat and on your having the right gut buddies to digest that food for you. For example, curcumin, the polyphenol in turmeric, increases the activity and expression of IAP and leads to a decrease in LPS production, inflammation, and intestinal barrier alterations. Other spices, including black pepper, red pepper, and ginger, also increase IAP levels.[23]

So it appears that polyphenols such as curcumin trigger IAP production, but remember what you learned earlier about polyphenols: we need our gut buddies to process them for us. This explains why probiotics increase IAP activity and reduce gut inflammation.[24,25] Getting more IAP requires us to nurture the right mix of gut buddies with the right foods. They, in turn, will nurture us!

Glycine

Glycine is the most important simple, nonessential amino acid in humans, animals, and many mammals. It is a precursor of several key metabolites, including, and importantly, glutathione. As I mentioned earlier, glutathione is one of only two antioxidants that work in the mitochondria. It's no surprise, then, that glycine, too, works directly on mitochondria, reduces inflammation, and helps maintain the gut lining. It also helps modulate the immune system, sending the signal that everything is fine and there's no need to attack with inflammatory cytokines.[26]

In fact, surgeons use glycine to suppress the immune system slightly during transplant surgeries. This leads to better results and lessens the likelihood of the body rejecting the new organ. Again, it's all about telling the immune system that everything is okay. It can lower its weapons and relax.

What does this have to do with the gut wall? Remember, everything that is good for the mitochondria is good for the gut wall. And although I'd hate to play favorites, the mitochondria in your gut wall are perhaps the most important ones in your body.

Further, there is evidence of exactly how important glycine is specifically for maintaining the gut wall. For instance, it is well known that the cells along the gut lining are susceptible to the toxic effects of radiation therapy, which disrupts the gut lining, leading to increased intestinal permeability.[27] The results are diarrhea, abdominal pain, nausea, and even sepsis.[28] In rats, glycine supplements protect the gut wall from radiation damage; its thickness, volume, and impermeability are all maintained.[29]

Glycine levels typically decrease with age, and this leads to elevated oxidative stress, glutathione deficiency, mitochondrial dysfunction, and increased inflammation. This comes as no surprise to me, since we know that glycine protects the gut wall and leaky gut is behind all of the diseases we normally associate with aging. However, we know that supplementing with glycine can be very effective. In one study, older adults who had severely elevated inflammation levels and endothelial dysfunction saw great improvements with glycine supplementation. After just two weeks, their mitochondrial function improved and they had a significant reduction in inflammatory markers.[30,31] Further, glycine supplements fully corrected mitochondrial dysfunction in patients with HIV.[32]

Glycine is also a major component of collagen. Perhaps this is why collagen can help maintain, reinforce, and repair the gut wall by modulating immune reactions, reducing oxidative stress, and helping distribute and express tight-junction proteins.[33]

Want one more benefit of more glycine? Taken as a supplement before bed, glycine lowers body temperature and induces high-quality sleep.[34]

Polyamines

Polyamines are organic compounds that have many jobs in your body, from aiding in cell proliferation and differentiation to acting as antioxidants, scavenging the body to eliminate free radicals.[35] There are three polyamines: spermine, spermidine, and putrescine. (Spoiler alert: the last one doesn't smell great.) Interestingly, your cells can make polyamines from amino acids, but your gut buddies actually make the majority of them in the large intestine.[36]

During times of rapid cell growth, especially the early stages of life, the presence of polyamines is especially important, and they are plentiful in breast milk.[37] They play a role in developing the inner ecosystem, particularly along the gut wall. They help the cells along the gut wall proliferate and differentiate and simultaneously aid in developing the immune system.[38] By helping to shape the immune system and acting as antioxidants, polyamines play a big role in modulating the inflammatory immune response.[39]

By helping cells grow and differentiate, polyamines help maintain the gut lining throughout your life. In mice, oral administration of polyamines helps repair the intestinal mucosa and reduce inflammation. It also increases the activity of intestinal alkaline phosphatase (IAP), another one of our other favorite gut wall defense mechanisms that you learned about above,[40] and increases antibody levels in the intestinal crypts to protect the precious stem cells.[41]

How can we get more of these goodies? Remember, the important compounds that are produced by our gut buddies can also be found in fermented foods. No surprise, this is also the case for polyamines, which are present in all sorts of fermented foods.[42]

This brings me back to the point that I'll keep making over and over throughout this book at the risk of being downright repetitive or evangelical: what's good for the goose is good for the gander. When your gut buddies are protected and well nourished, they ferment essential compounds that help keep you healthy and that protect themselves, too, so they can continue to take good care of you. But it must all start with a healthy, protected, well-functioning ecosystem. It's as if your gut buddies are saying "Help me help you."

The problem, of course, is that we have not helped them. And that is exactly why they are no longer helping you. It's not that they don't want to; it's that they literally can't because they have been ignored, killed off, mistreated, and left defenseless. This brings me to the next chapter, in which you'll learn about all of the ways that we've so badly damaged our inner ecosystem. Then I'll dive deeper into exactly how this has manifested itself in nearly every common disease—and, most important, how we can prevent and heal them.

CHAPTER 4

A PERFECT STORM

In the classic film *The Graduate*, a middle-aged man approaches Benjamin Braddock, played by the young Dustin Hoffman, at a party and says, "I just want to say one word to you. Just one word." After a dramatic pause, he proclaims, "Plastics! There's a great future in plastics!"

He was right. *The Graduate* was made in 1967, and in the fifty-plus years since then, the rate of plastic production has grown more than that of any other material. As consumers, we love plastic. It's cheap, lightweight, convenient, and versatile. And of course, the plastics industry loves it, too. The global plastics market is now valued at $609 billion, and that number is expected to go up to more than $811 billion by the year 2030.[1]

Plastics are ubiquitous. They're everywhere, from our landfills to our oceans to the foods we eat, the water we drink, and even the air we breathe. But the bright future of plastics has arrived with a major downside. Besides its enormous environmental impact, it is devastating to our internal ecosystem. It is toxic to the microbiome. Read that again: toxic. It negatively affects microbial composition and diversity. These changes alter the hormone balance and immune response, creating the ideal environment for disease.[2]

But the problem is far bigger than just plastics. Over the past fifty years, the increase in plastics production has coincided with the advent of other technologies, medicines, and trends—from scented candles to "whole grain goodness"—that have wreaked havoc on our inner ecosystems. Ironically, some of these, such as nonsteroidal anti-inflammatory drugs (NSAIDs) and proton pump inhibitors (aka antacids), are taken to ease the symptoms of an imbalanced microbiome but only serve to exacerbate the problem. This creates a terrible cycle of pain and sickness, harmful medications, more pain and sickness, and ever more harmful medications!

At the same time, our diets have changed radically, becoming increasingly devoid of the foods our gut buddies need to live and thrive in harmony. Even worse, we've replaced them with artificial, processed, chemical-laden foods that literally starve our gut buddies to death while allowing the bad guys to reproduce and flourish. And I haven't even touched on broad-spectrum antibiotics that wipe out our microbiomes or the harmful chemicals that are now in the air, in our water supply, and sprayed directly onto our food.

Altogether, we have created the perfect storm for a dysfunctional, dysbiotic, and downright decimated rain forest. We truly could not have designed a better system to destroy our microbiome if we'd tried.

It's fair to say that we are now paying the price. As I'm sure you know, over the past fifty years there has been a dramatic uptick in nearly all chronic diseases. Since the 1970s, obesity rates have consistently increased among adults and have more than tripled among children. Today, 42 percent of American adults are obese. This has, of course, led to an increase in type 2 diabetes, and the American Heart Association predicts that 45 percent of people in the United States will have heart disease by the year 2035. Meanwhile, early-onset cancer (in patients under the age of fifty) is dramatically increasing around the globe, with the United States leading the way. Someone in the United States develops Alzheimer's disease every sixty seconds, and Parkinson's disease is increasing even more rapidly; it is the fastest-growing neurological disorder in the world.[3]

We are also in the midst of multiple epidemics. The National Alliance on Mental Illness reports that rates of mental illness have been increasing year after year, beginning long before the covid pandemic damaged so many people's mental health. And over the past twenty-five years, there has been a 44 percent increase in the types of antibodies present in lupus and other autoimmune diseases—diseases that used to be extremely rare. Well, they are certainly rare no longer. I would argue that every one of the diseases I've mentioned so far is related to autoimmunity, leaky gut, and intestinal dysbiosis. In other words, they all stem directly from the gut.

A great future for plastics, indeed. A less rosy future for the environments both inside and outside us.

The main thing that I want you to remember as you read this chapter is that no matter how badly you may have unwittingly damaged your microbiome, it's not your fault. We have all been sold a false bill of goods by leaders and so-called experts who had no idea about our microbiome—even the fact that it existed—or how many "lifestyle-enhancing" objects could hurt it. Once they did know, they failed to share the information with the public, again and again choosing profits over your health. You did the best that you could at the time based on the information you had, and now that you know more, you can do better.

Most important, it's not too late. We can prevent, halt, and reverse the vast majority of these diseases by eliminating gut destroyers and restoring our gut health. First, let's take a closer look at what has created this perfect storm and how it's affecting us.

BROAD-SPECTRUM ANTIBIOTICS

I've discussed some specific types of gut buddies, but I haven't yet explained the two main categories of bacteria: gram negative and gram positive. Back in 1884, a Danish bacteriologist named Hans Christian Gram devised a test to determine whether bacteria had thick or thin membranes. Those with thick membranes are referred to as gram positive, and those with thin membranes are gram negative. Gram-negative bacteria also have an outer membrane containing lipopolysaccharides—that's right, those pesky LPSs! In both types of bacteria, the membranes protect the bacteria while allowing nutrients in and keeping waste products out.

I'm explaining this because it's important to understand in the context of antibiotics. The first antibiotics that were developed could kill only one of these two types of bacteria. For instance, penicillin killed primarily gram-positive bacteria, whereas streptomycin was active against gram-negative bacteria. Doctors prescribed one of them based on what type of infection a patient had. Of course, the drugs still impacted the gut biome, but, remember, an infection was present. One type of bug was overgrowing, causing an infection, and those targeted antibiotics put a stop to it. For the most part, they left the gut buddies that were just minding their business alone, leaving the microbiome relatively intact.

Then, in the late 1960s, the first broad-spectrum antibiotic was developed. "Broad-spectrum" means that it kills both gram-positive and gram-negative bacteria. That made things easier for doctors because they didn't have to bother figuring out what kind of infection a patient had. They could just prescribe the new drug, and—presto!—no more infection.

The problem with these drugs, however, is the exact same feature that makes them so appealing. They kill a *broad spectrum* of the microbiome: the good, the bad, and the ugly. This wipes out the microbiome, transforming the teeming rain forest into a desert wasteland.

But it gets worse. As broad-spectrum antibiotics gained popularity over drugs such as penicillin in the mid-1970s, which just so happened to be in the same fifty or so years we're discussing, they also started being overprescribed. It was so easy for doctors to give them to patients who had any kind of infection and get quick results that they started to give them to patients with any *sign* of infection. But many symptoms of infection mirror those of common viruses, which are usually resolved by a healthy immune system with no outside intervention. Today, a full 25 percent of antibiotic prescriptions are considered inappropriate or unnecessary, and the average American is prescribed a course of antibiotics every six months.[4]

This doesn't affect just the individuals taking the drugs, though it surely does that. I think that by now we've established exactly how intelligent bacteria are. They do not remain static over time. They have coevolved with us over trillions of years, and they continue to do so. All beings, as individuals and full species, must evolve in order to adapt and survive in a changing environment. And bacteria have quickly evolved to become immune to antibiotics so they cannot be killed off so easily.

This has led to an arms war between scientists and bacteria. As new drugs are developed, bacteria adapt and become stronger. But where will this lead? Even today, mortality rates are high among patients with antibiotic-resistant infections,[5] and common diseases such as pneumonia, tuberculosis, and salmonella are becoming more difficult to treat, as the antibiotics that were usually used against them are becoming less effective. Antibiotic resistance is now considered a public health crisis.[6]

Plus, it's not only humans who are given antibiotics. Livestock are commonly treated with antibiotics even when the animals do not have active infections. This is for two reasons that have nothing to do with health—yours or the animals'—and everything to do with profits. One, antibiotics make animals gain weight more quickly. (And guess what? They do the same thing to you.) Two, most livestock suffer from such inhumane living conditions that infections are unfortunately very common. Infected animals mean less money, so antibiotics are given as a preventative, prophylactic measure. These drugs are present in the products you eat from treated animals,[7] essentially providing you with a low dose of antibiotics with every portion. This contributes to the antibiotic resistance crisis and, more important for you today, it damages your microbiome.

How exactly does all of this damage your microbiome? Just one course of antibiotics leads to a profound, rapid reduction in microbiome richness and diversity,[8] which, you will recall, are the hallmarks of a healthy microbiome. This sets the stage for opportunistic bad guys to take over, making you more susceptible to infections such as *Clostridioides difficile* (*C. diff.*), which causes diarrhea and colitis and can even be deadly.[9]

Your gut buddies want to live in a state of homeostasis, and they try to get back into balance after a course of antibiotics, but their restoration efforts are often incomplete. The exact changes that remain in place and the extent to which the microbiome remains disturbed is highly individual among patients. In some cases, the microbiome regains stability but with a different makeup than it had before the antibiotics.[10]

Let me return to my forest fire example from the last chapter to explain this phenomenon. A forest is a complex ecosystem that develops over an extended period of time, with interdependent flora and fauna. When it burns to the ground, even though thousands of seedling trees are planted afterward, it will take twenty to thirty years before the complex ecosystem is reestablished and the forest returns to normal.

The same thing happens in your gut ecosystem rain forest. Sure, you can swallow some probiotics (friendly bacteria), but how naive we are to believe that we can rebuild that complex ecosystem presto-changeo! In fact, numerous studies on healthy human volunteers given the broad-spectrum antibiotics that are commonly prescribed for upper respiratory infections (which are usually caused by viruses, not bacteria) showed huge perturbations and decimation of the microbiome lasting from two months to as long as two years. Even after recovery, the microbiome looked profoundly different than before, with a lot of gut buddies completely missing.[11,12]

Well, you ask, so what? These changes can be very damaging. There is a clear, well-established link between antibiotic use and obesity[13]—remember those nicely fattened-up livestock?—and other diseases. Mice that were treated with antibiotics spontaneously developed type 1 diabetes,[14] had increased cytokines in their colon, and showed decreased levels of SCFAs and other important metabolites produced by bacteria.[15]

Without these signaling devices, communication between your gut buddies and mitochondria is cut off.[16] This has dire consequences for your immune system and every cell in your body. Without these metabolites, the mitochondria do not know when to uncouple and reproduce or when or how to die; your immune system does not know when and where or what to attack; and the stem cells in your crypts don't know when they are needed to fortify the gut wall—and that's just a start. It's no surprise, then, that antibiotic usage is linked to increased oxidative tissue damage,[17] suppressed gene expression, and damage to the mitochondria.[18]

This all happens in adults, yet antibiotics are possibly even more damaging when taken during pregnancy and early childhood. When an expecting mother takes antibiotics that alter her microbiome, it leads to a decrease in the fetus's microbiome diversity,[19] reduced immunity,[20] and a likelihood of disease later in life.[21] Infants who were exposed to antibiotics in the womb have an increased risk of developing colitis,[22] cerebral palsy, epilepsy, and cardiac and genital malformations.[23] They are also more likely to develop asthma, allergies,[24] functional impairments in development and cognition,[25] obesity,[26] and diabetes.[27]

Antibiotics are most damaging to the fetus when taken by the mother during the first trimester.[28] Additional rare birth defects such as microphthalmia (a severe eye defect), hypoplastic left-heart syndrome (when the left side of the heart doesn't develop correctly, restricting blood flow), atrial septal defects (a hole in the heart between chambers), and cleft lip with cleft palate are also associated with the mother's use of antibiotics usage during the first trimester.[29]

After birth, mothers who take antibiotics while nursing have altered milk microbiota, which influences their baby's developing microbiome.[30] Of course, infants and children who take antibiotics at the critical stage when their microbiomes and immune systems are developing have altered microbiomes, as well.[31] This leads to a higher risk of developing asthma,[32] allergies,[33] and obesity[34] later in life.

Let me be clear: I am not advising against antibiotics or any other medication when they are prescribed for life-threatening infections and taken responsibly. But antibiotics have been mismanaged and overprescribed, causing world- and life-altering side effects. It is essential to take or give your children antibiotics only if and when they are absolutely necessary.

However, if you do need to take antibiotics or have taken a lot of them in the past, please don't worry that you're now destined to suffer from dysbiosis and disease. By following the *Gut Check* program before, during, and/or after a course of antibiotics, you can protect and restore your all-important inner ecosystem.

GLYPHOSATE

While we have been destroying our inner ecosystems by abusing oral antibiotics, we have also been spraying them on our food. Literally.

Glyphosate, the herbicide in the weed killer Roundup, was patented as an antibiotic. It works by disrupting the pathway that plants use to produce amino acids, called the shikimate pathway. Plants but not animals use the shikimate pathway. That led scientists to assume that glyphosate would kill weeds while remaining safe for humans and other animals. Makes sense, right?

Well, not so fast. Guess who else uses the shikimate pathway? Your gut buddies, of course. Glyphosate is lethal for them, which means that it's big trouble for you.

I've written about glyphosate in some of my previous books, but there is now new information about and evidence of exactly how damaging it is to the microbiome and therefore to us. Glyphosate was first introduced to the market in 1974. Immediately, plants, in all of their wisdom, found a way to evolve and become resistant to glyphosate, just as bacteria have with antibiotics. By the early 1990s, glyphosate-resistant crops led to a massive increase in glyphosate use. *Hey, this stuff isn't working. Let's use more of it!* Not the smartest idea on the part of humans. If plants can evolve to be resistant to a little glyphosate, what did we think was going to happen when we started applying a lot more? Apparently, no one stopped to consider that, and by 2019 US farmers were spraying more than 280 million pounds of glyphosate on their crops each year.[35] This increase in glyphosate usage has obviously led to more and more glyphosate residue in our food.[36]

In the years since then, we've seen evidence that glyphosate is dangerous and carcinogenic,[37] but we didn't know exactly how it affected the gut. Then, in 2021, scientists at King's College London conducted a groundbreaking study that looked at the effects of glyphosate on the infant microbiome.[38] They found that although maintaining pH levels is key to gut homeostasis, glyphosate creates an acidic environment that is deadly for some of our gut buddies and allows others to overgrow.

Not only does glyphosate destroy the state of homeostasis, but unfortunately, some of the bacteria that are the most naturally resistant to glyphosate are also the most harmful. They have the potential to increase the production of pro-inflammatory cytokines and reactive oxygen species (ROSs). Compounding this problem, even low-dose glyphosate exposure reduces the key enzymes that we need to reduce ROSs, particularly superoxide dismutase and glutathione reductase.[39,40] That's a double whammy in the fight against cellular damage.

Even worse, some of our gut buddies that are most susceptible to glyphosate are also the most essential. These include *Lactobacillus*, Ruminococcaceae, and *Butyricicoccus*, which are key producers of SCFAs, tryptophan, and L-glutamate. You already know how important SCFAs are. Tryptophan and L-glutamate are also essential, particularly when it comes to your mental health. These amino acids are precursors of many important neurotransmitters, including serotonin, dopamine, gamma-aminobutyric acid (GABA), and epinephrine.[41,42] These are all mood-stabilizing chemicals that our bodies cannot produce without the help of our gut buddies. More on this later.

Furthermore, your gut buddies normally metabolize tryptophan into additional compounds, including 3-indolepropionic acid (IPA), an important antioxidant in the brain that decreases the activation of glial cells, the brain's immune cells. This protects against DNA damage and helps prevent Alzheimer's disease.[43] It should come as no surprise, then, that a reduction in the gut buddies that are most susceptible to glyphosate is implicated in multiple neurological diseases, including Parkinson's disease,[44] schizophrenia,[45] and depression.[46]

As I said, I've been writing about the dangers of glyphosate for years, but the evidence against it keeps compounding. So why are we still using it? Back in 2017, leaked documents from Monsanto (the company, now owned by Bayer, that sells and profits from Roundup) showed that it had created its own research to "prove" that its product was safe while covering up internal disagreements about its safety. Yet it still has not been banned in the United States. Two years later, in 2019, the Environmental Working Group found glyphosate in all twenty-one popular breakfast cereals that it tested, including several organic varieties.

Meanwhile, many European countries have already banned glyphosate. Those that do allow it have lower limits than the United States does. It's no coincidence that so many of my patients travel to Europe and feel great after eating the bread and pasta that make them sick at home. They come back and tell me, "Dr. G., I'm cured!" Sadly, when they extend their

European eating habits to the United States, they often end up falling sick again. Not only do our taste buds know the difference when they get a bite of the "good stuff," but our gut buddies know it, too.

ENDOCRINE DISRUPTERS

Scientists have known for a long time that certain chemicals in our environment mimic the behavior of hormones in our bodies, throwing our endocrine systems out of whack. These are called endocrine disrupters, and they include pesticides, bisphenol A (BPA) (found in canned foods and plastics that were made before 2012), phthalates (found in plastics and fragranced products), parabens (found in many processed foods and personal care products), heavy metals (found in many cosmetics and in our food and water supply), and oxybenzone (found in sunscreens). What we are just now learning, however, is the connection between these endocrine disrupters and our microbiome.

The first and most impactful way that endocrine disrupters affect us is obviously through our endocrine and hormonal systems. In males, endocrine disrupters affect sperm production and quality, and in females, they affect conception, pregnancy, and delivery.[47] They do this by interacting with hormone receptors by turning them on or off and by hijacking the signals that are sent to those receptors.

Endocrine disrupters can also impact circulating hormone levels in the body. They do this by altering the permeability of the cell membranes that allow hormones in and out. In other words, these chemicals direct where and when your hormones should go and how the rest of your body should respond to the fluctuating hormone levels. The impact that this has on your body is enormous. Even mild hormone imbalances can lead to everything from infertility to anxiety, depression, bone disease, heart disease, and cancer.[48]

You may already be thinking, *Hey, that whole signaling thing sounds sort of similar to what our gut buddies do!* If so, you are correct. Endocrine disrupters make a lot of their mischief by messing with your gut buddies' signaling molecules—those uber-important communication tools. For one, they interfere with the way short-chain fatty acids (SCFAs) signal and modulate glucose metabolism. We now believe that this is the connection between endocrine disrupters and the increased risk of diabetes. They also interrupt the important signals being sent by neurotransmitters, cytokines, and hormones.

Endocrine disrupters can also create changes to microbiome composition, leading to dysbiosis. Certain types of gut buddies are more vulnerable than others to specific chemicals, leading to uniquely dysbiotic makeups. For example, BPA creates changes in the gut similar to those of a high-fat diet![49] There go the benefits of your healthy eating. And before you say that you don't use BPA plastic bottles anymore, BPA replacements, such as BPS and BPF, are no safer and have been recently linked to the causation of hormonal cancer.[50]

We are especially vulnerable to the changes elicited by endocrine disrupters in the womb and during early infancy, when gut homeostasis is first established. But exposure to endocrine disrupters can create dysbiosis at any time. In 2019, French scientists found that in aquatic organisms, exposure to these chemicals led to an increase in bad guys in the gut and a decrease in friendly gut buddies.[51] In humans, endocrine disrupters are associated with polycystic ovarian syndrome (PCOS), diabetes, allergies, asthma, and autoimmune diseases.[52] They also lead to decreased expression of tight-junction proteins, impairing the gut wall and leading to leaky gut.[53]

This explains why some people are now saying that the use of fragrance is the new smoking. Endocrine disrupters are in the synthetic fragrances in everything from air fresheners and candles to perfumes. As a little tease of something that I'll talk about in much greater detail later, your gut buddies would much rather you smoke than wear perfume. In fact, I think they'd rather enjoy it.

WHOLE GRAIN GOODNESS

It wasn't until 1994 that the US Department of Agriculture (USDA) started defining for us what was "healthy" to eat, and it was shortly after that, with the support of the Whole Grains Council, that it started encouraging Americans to eat more whole grains. Now, remember that the USDA's job is to promote agriculture, not health, yet it is in charge of creating our "healthy" food pyramid. That's the equivalent of putting drug companies in charge of most of the research and education of physicians in the United States, which would be a profound conflict of interest. But wait! That's exactly what has been done both with drugs and with food. The fox is clearly guarding the henhouse.

Of course, historically, humans never needed a government agency to tell them what to eat. And when left to their own devices, our ancestors brilliantly chose exactly what to eat and how to store and process those foods for the direct benefit of their gut buddies. That includes how they processed whole grains. Did the USDA ever stop to wonder why ancient civilizations always refined their grains through a process called milling? Doing so removes the bran, germ, and husk from the grains.

The ancient Romans are credited with being the first people to mill grains. The process required labor and materials and was expensive, so the use of refined grains became a status symbol. Only the wealthy elite ate white bread, while slaves and members of the lower classes consumed whole grains. That remained the case for much of human history. In a remarkably similar fashion that I refuse to believe is a coincidence, Asian cultures began removing the bran from rice thousands of years ago to create white rice. Until relatively recently, white rice was seen as a luxury that was reserved for society's elite.

This poses the million-dollar question: If the process of refining grains was so expensive and labor intensive, why did all of those people bother to do it? I am certain that it was for the same reason that our ancestors fought wars over spices and fermented their foods—because they knew on an instinctual level that those foods were healthier.

Of course, with modern scientific techniques and measurements, now we know exactly why these foods are healthier—because they are better for our gut buddies. That's right, despite everything that we've been told about how badly we need whole grains and their fiber and nutrients, refined grains—and, no, that doesn't mean highly processed!—are much, much better for our gut.

So what's the problem with whole grains? In a word, lectins. One type of lectin that we are learning more about every day is wheat germ agglutinin (WGA). As the name implies, WGA is part of the wheat germ, which is removed during milling. Why bother removing it? Because WGA binds to sialic acid, part of the sugary coating on so many of our cellular surfaces, including blood vessels, the synovial surfaces of joints, the nearly impermeable blood-brain barrier, the myelin sheath that encases and insulates your nerves, and even the surface of your eyeballs. This coating is called the glycocalyx. Like the mucus layer of the gut lining, these sticky surfaces are meant to trap invaders and keep them away from your delicate blood vessels, nerves, joints, and other critical parts.

This is important because WGA is an invader and acts on the glycocalyx as a splinter does on your finger. The problem is, this small lectin can leak through the gut barrier even if you don't have leaky gut.[54] And when you regularly consume whole grains, as the USDA wants you to, this happens over and over throughout your body. Of course, this causes your immune system to become overactivated, leading to inflammation and disease.[55]

A glycocalyx protects many, many parts of our bodies, and WGA can bind to each of them, prompting an autoimmune attack on those tissues. We have seen this in everything from the vascular endothelial cells that line our arteries, veins, and capillaries[56] to the glycocalyx lining our eyes.[57] Furthermore, because it is so small, WGA can penetrate the blood-brain barrier,[58] a layer of endothelial cells that protects our brains from invaders. This leads to inflammation in the nervous system.

In the next chapter, I'll share more details about how this leads to disease, but I'll bet that you are already starting to get the picture. For now, we must understand how harmful the last several decades of "whole grain goodness" have been to our gut and nearly every system in our body.

NONSTEROIDAL ANTI-INFLAMMATORY DRUGS (NSAIDS) AND PROTON PUMP INHIBITORS (PPIS)

I've talked a lot about the importance of uncoupling mitochondria, but we all know that there can always be too much of a good thing. For some background, I'll share a brief story that will be familiar if you read my book *Unlocking the Keto Code*. During World War I, many workers at munitions factories in France and Germany began rapidly losing weight. Researchers soon found out that 2,4-dinitrophenol (DNP), a compound used to make explosives, was responsible. Soon, it was packaged and sold as a weight loss drug. DNP increased users' basal metabolic rate, causing them to burn more calories from food and from their fat stores, leading to drastic weight loss.

Sounds pretty good, except for the fact that the people who took it started suffering from many side effects, including fever, cataracts, thyroid issues, and death. DNP was soon banned and mostly forgotten. Roughly forty years later, in the 1970s, it was discovered that it caused weight loss through rapid mitochondrial uncoupling. But it did so without the benefit of mitogenesis.

Remember, uncoupling works well when other mitochondria are created at the same time to share the workload. When one mitochondrion uncouples without the birth of more mitochondria to share the burden, it produces less ATP. This can lead to cell death and damaged tissue. Think back to my dogsled analogy in chapter 2. This is akin to having only one dog that is exhausted pulling the sled, leaving you going nowhere fast. All of the side effects attributable to DNP were the results of cells' dying from a lack of ATP.

What does this have to do with NSAIDs? To a lesser extent, they do the same thing as DNP. When they are absorbed into the gut wall, they cause a drastic uncoupling of the mitochondria in the wall of your gut without stimulating the birth of new mitochondria via mitogenesis.[59] The gut lining cells die from a lack of ATP, and you are left with holes in your gut wall—good old leaky gut.

NSAIDs cause dysbiosis along with bleeding, inflammation, and ulceration in the stomach and small intestine.[60] Within just one hour of an NSAID being taken, it creates dose-dependent changes to mitochondria.[61] It should come as no surprise, then, that long-term NSAID users have an increased risk of heart disease and stroke.[62]

Furthermore, NSAIDs actually exacerbate inflammation, and not just in the gut. A new study out of the University of California shows that after one year of NSAID treatment for osteoarthritis, patients had the same levels of inflammation. After four years of treatment, their inflammation was actually worse.[63] Of course, when patients continue to complain of pain after NSAID treatment, they are often prescribed—you guessed it—more NSAIDs!

If the symptoms caused by the NSAIDs are in the upper GI tract, doctors often prescribe proton pump inhibitors (PPIs), which reduce stomach acid, but these drugs only cause further damage. Like most things in your body, stomach acid is there for a reason—in this case, more than one reason. Stomach acid is meant to destroy lectins and protect you from bacteria that shouldn't get past the stomach. Remember, an acidic environment is deadly to bacteria.

As acids move down the GI tract and the inner ecosystem becomes more and more densely populated, they are naturally diluted by the addition of pancreatic juices and bile. By the time the end of the small intestine is reached, the acid is mostly neutralized. This is called the acid gradient, and it keeps bacteria from climbing up higher into the intestine.

If PPIs are used to get rid of the acid, there goes the gradient, and there's nothing stopping bacteria from swimming around into the small intestine, where they don't belong. This can lead to small intestinal bacterial overgrowth (SIBO), which often creates problems with immunity.[64] PPI users also have a significantly reduced abundance and diversity of gut buddies in the lower GI tract[65] and an increased risk of infections such as *C. diff.*[66]

But hang on a sec, what do protons have to do with any of this? Of course, we already know that protons are used to make ATP by pumping them from the outer to inner mitochondrial membrane. It turns out that we produce both ATP and stomach acid by pumping protons. When developing PPIs, scientists considered only half of this equation and assumed that it would be a good idea to inhibit the proton pumps in the stomach. But this simultaneously keeps mitochondria from producing ATP. Whoops.

It is now widely known that these drugs are not safe to take for long periods of time. It says right on the box that you shouldn't take them for more than two weeks. Yet I see patients who have been on them for decades. Meanwhile, long-term use of PPIs is associated with an increased risk of developing dementia,[67] pneumonia,[68] kidney disease,[69] heart disease,[70] and reduced bone density.[71]

This is, sadly, what happens when scientists look at the human body through a narrow lens. Your gut buddies and the inner workings of your gut in general are a beautiful example of how everything is connected and, when left interrupted, works together in perfect harmony. When one element of the system is disrupted, it all falls apart.

MICROPLASTICS

Let's wrap up the description of the assault on our gut buddies right where we started: with plastics, particularly microplastics. Though some damaging compounds such as BPA have been banned from plastics, plastic particles called microplastics, which are a mere micrometer in size, are now omnipresent in the environment. These are both manufactured directly and released into the environment as fragments of larger plastics. They are in our air and our food chain, inhaled and consumed by humans every day.[72]

In fact, a recent study revealed up to twenty different microplastics in the stools of eight healthy volunteers, seven of whom were eating fish.[73] You might assume that this is no problem because they pooped them out, right? Nope, sorry. Their effects are far-reaching and, quite frankly, scary.[74]

Microplastics alter microbiome diversity, create dysbiosis,[75] and impact gut barrier function.[76] Exposure increases the abundance of bacteria that are associated with respiratory diseases (*Klebsiella* and *Helicobacter*) and digestive tract diseases (*Bifidobacterium*, *Streptococcus*, and *Sphingomonas*) and reduces the abundance of beneficial bacteria, such as *Bacteroides*, *Ruminococcus* torques group, *Dorea*, *Fusobacterium*, and *Coprococcus*.[77] And when microplastics are ingested, they can accumulate in the intestines, injuring epithelial cells and creating toxic effects.[78] Worse yet, they cross the placenta into the bodies of unborn babies.[79]

The people whom the government have put in charge of our health seem to believe that microplastics are too small to make a significant difference. But as we've now seen, it is the little things that truly have the biggest impact on our health and well-being. Yet this is the story of Western civilization: we have ignored the small things that are the most important, denied them what they want, and aggressively tried to kill them off.

Suffice it to say that they're mad as hell and they're not going to take it anymore.

CHAPTER 5

HIPPOCRATES WAS RIGHT

Nearly every day, as I'm either doing research or treating my patients, I have the same two thoughts: One, *How did Hippocrates know?* He lived twenty-five centuries ago. To give you some context, at that time, the human life expectancy was thirty-five years. Humans believed that the earth was flat, had no concept of evolution, and were unaware of the fact that there were other planets in our solar system or that dinosaurs had existed. Yet Hippocrates somehow knew a simple, profound, and mysterious truth that we are just beginning to see the full evidence of today: *all disease begins in the gut*.

To take it a step further, death itself begins in the gut, as the microbiome is also the root of aging. *C. elegans* (roundworms) are common and useful models for study since their genome has functional counterparts in humans. In the *C. elegans* model of aging, we clearly see that as the intestinal wall breaks down, so does the worm.[1] The same is true in humans.

This goes hand in hand with the second thought that so often crosses my mind: *How did we get it so wrong?* Not only have we destroyed our critical ecosystems in all of the ways I discussed in the previous chapter, but we've also ignored even the most obvious connections between every major disease and the gut. Though Hippocrates had no material evidence back when he made his statement, for many years now we have been confronted with overwhelming data supporting his theory. Yet for the most part, we haven't paid any attention and have too often gone in the exact opposite direction when diagnosing and treating diseases, looking at the symptoms and not the root cause.

As you're about to see, gut dysbiosis and leaky gut are implicated in every major disease, and they have been for a long time. Yet even researchers who saw the connection between diseases and the gut so often interpreted it backward. When they observed the fact that patients with cancer, heart disease, diabetes, and even bone and joint issues *happened* also to have gut dysbiosis and leaky gut, they assumed that the disease states were causing problems with the patients' guts. In fact, it's well known that Parkinson's patients have problems with constipation, so until very recently it was assumed that the Parkinson's disease caused the constipation. (More on this in a moment.) But there was no evidence of or even theories as to how diseases could lead to problems in the gut.

Well, of course there weren't, because it doesn't work that way! In fact, it is the exact opposite. The patients all had leaky gut and dysbiosis because those were at the root of their disease, not the other way around. I would go so far as to say that all of the major diseases I mentioned are in fact the same disease—leaky gut—and what we refer to as a disease is merely the manifestation of the symptoms caused by leaky gut.

MOLECULAR MIMICRY AND AUTOIMMUNITY

I've already covered how leaky gut works and why it's so harmful. In brief, a dysfunctional gut wall allows invaders such as lectins, LPSs, undigested "normal" food particles, and bad bugs to sneak through the gut barrier and into your bloodstream, where they do not belong. The TLRs on your immune cells recognize the invaders as foreign and respond by sending a signal to the rest of your immune system, telling it to stay on high alert and launch an attack. The result is inflammation, that old "root cause" of disease.

As a reminder, short-term inflammation isn't a bad thing. Inflammation is part of the immune response for a reason: it's a result of the flood of white blood cells swooping into the site of an infection or damage to help you heal. There are certainly times when you need that. The problem is that when you have leaky gut and your body is constantly responding to invaders, your body becomes perpetually inflamed. This sets the stage for disease.

Another important part of this puzzle is the fact that many of the proteins that slip through your gut wall and latch onto your body bear a striking resemblance to your own tissues. My friend and colleague Professor Loren Cordain at Colorado State University first coined the phrase "molecular mimicry" to describe this phenomenon. When the immune system is called on over and over to fight off a foreign protein, it becomes hyperactivated and starts attacking anything that remotely resembles that protein—namely, human tissues. It is literally a case of mistaken identity.

Here's how I explain this idea to my patients: Plants don't want you to eat them or their seeds, which are their babies. So they incorporate defensive proteins called lectins. As you learned earlier, if these lectins get past your defenses of stomach acid, the microbiome, and the mucus, they can attach to your gut wall, flip a few switches, and create gaps in the wall that they can pass through.

A huge part of your immune system lies in wait just across this border, ready not only to attack locally but also to send out a total body alert via cytokines that you are being attacked. Your body now scrambles the fighter jets (other immune cells) armed with pictures of the rogue protein bar codes that have invaded and begins searching your entire body for the troublemakers.

I think that plants are amazingly clever and have intentionally made a number of their protein bar codes look similar to those of other proteins in your body. Let's take an autoimmune disease such as Hashimoto's thyroiditis as an example. You have leaky gut, so the fighter jets have been scrambled and are going around your body looking for proteins that have a bar code they have been sent to identify and attack. They fly past your thyroid, and, lo and behold, there are bar codes (proteins) in your thyroid that look remarkably similar to what they are looking for! They are not quite the same, but hey, we are at war, and because they look so similar, it's better to shoot first and ask questions later. And boom!

You attack your own thyroid and now have an "autoimmune disease," all as a result of "molecular mimicry." Or, rather, all as a result of leaky gut.

Hold that thought for a moment, and consider what may not be obvious at first glance. What makes the fighter pilot (your immune system) so willing to pull the trigger and shoot at something it knows isn't quite the same protein that it's looking for? Why is it so trigger-happy?

I can assure you that your gut barrier, even with a perfect defense system, is breached every so often. That's why so much of your immune system lines your gut. Since this should be only an occasional occurrence, most of the time your immune system sits around having a cigarette and a doughnut and taking it easy. "Nothing to see here, folks, move along."

Contrast that with a gut wall that is constantly being torn open, with lots of foreign proteins and bacteria coming across the border. Now, instead of relaxing, the immune cells are hypervigilant. And rather than being armed with nightsticks, they are now carrying AK-47s, multiple magazines of ammo, and a Kevlar vest and are truly "locked and loaded." Previously, when there wasn't much going on, a suspicious-looking protein in your thyroid would probably have been met with a shrug. But now that same protein in your thyroid will be blown to smithereens because of that itchy trigger finger!

This is a big part of why lectins in particular are so damaging. Lectins and the proteins on and in many of our organs, nerves, blood vessels, and joints share similar molecular patterns. So when lectins are constantly seeping through a leaky gut (or when you consume WGA, which is small enough to get through the gut wall even if it's intact!) and you have LPSs constantly putting your immune system on high alert, your immune system begins to view parts of your own body as these invaders and attacks them, too. When the immune system attacks the body, it's called an autoimmune disease.

I successfully treat these diseases in my clinics using the *Gut Check* program every day and now have upward of five thousand autoimmune patients who are no longer suffering from their disease. How do I do this? It's not magic. Once a leaky gut is sealed, the immune system begins to calm down and is no longer on guard against those particular invaders. So it stops attacking anything that looks like them.

Of course, as part of the *Gut Check* program, I also work with my patients to "reseed" their microbiomes to restore a robust, homeostatic ecosystem. And boy, does it work! As I reported at the American Heart Association in 2018, 94 percent of patients with autoimmune diseases (as proven by blood markers) were cured or in remission within nine months of following my *Gut Check* program.[2]

One autoimmune condition that I often treat is psoriasis. About five years ago, a patient came to me as a seven-year-old kid who couldn't walk because of weeping sores on his hands and feet. He had been to many specialists around the world, who had offered little help. Yet we know that the differences in the microbiome in patients with psoriasis and other people are profound and consistent.[3] Sure enough, after looking at the patient's blood work, I could see that he had leaky gut, which we were able to correct with a *Gut Check*. He is now a healthy twelve-year-old who plays sports and fights with his mother about following the rules.

Remember, the gut isn't the only home of bacteria in our bodies. As I mentioned earlier, our oral microbiome is also incredibly important. Our oral mucosa functions very much like the gut wall to keep bacteria out of our bloodstream. When the oral microbiome is out of balance and bad bugs are able to access our bloodstream that way, they can cause just as much damage as they do when they enter through the rest of your gut.

A recent study looked at the oral microbiomes of patients with rheumatoid arthritis, an autoimmune disease that affects millions of Americans.[4] It showed that patients with rheumatoid arthritis were more likely to have periodontal disease than healthy people were. In fact, patients with rheumatoid arthritis had high numbers of bacteria from their oral microbiome in their bloodstream! They also had high levels of inflammatory immune cells in their bloodstream during flares. All of these signs point to the "leaky mouth" and molecular mimicry theories of autoimmune disease.

Of course, the makeup of the microbiome is also incredibly important—and not just to maintain the gut wall. When you have a robust and balanced internal ecosystem, your gut buddies will work together to produce all of the necessary postbiotic messaging signals I discussed earlier. These messages keep the mitochondria and immune system up to date about how things are going down in the gut and contain instructions on when and where to attack and when to stand down, take a break, and relax.

By decimating our gut buddies, we've disrupted this vital communication system. When the immune system never receives the signal to relax, things can go downhill fast. Add leaky gut and chronic inflammation to this picture, and you might start to wonder how any of us has managed to avoid suffering from an autoimmune condition, a dis-ease! Well, unfortunately, folks, most of us haven't. And that's because this same mechanism is behind every major disease. It's really just a matter of where in your body the immune system erroneously attacks.

LEAKY GUT = LEAKY HEART

As I said earlier, lectins are sugar-seeking proteins. They attach to sugar molecules made of sialic acid in the gut lining. This is why you have the mucus layer: to trap and protect you from invaders such as lectins before they can bind to your gut wall and pop it open.

Well, guess what? Your gut lining isn't the only surface in your body that's lined with sugar molecules. You also have them along your blood vessels, your joints, the junctions between your nerves, the coating over your nerves, your eyeballs, and even, as I'll discuss later, in the blood-brain barrier that protects your brain. These coatings are collectively known as the glycocalyx.

One such glycocalyx is the thin layer of polysaccharides and lipids that forms a protective layer overlaying the surface of your blood vessels' endothelial cells. Its coating is similar to the mucus overlaying the endothelial layer of the gut. When you have a leaky gut that allows lectins to get through the gut lining, they can bind to and damage the lining.

So now there is a splinter stuck against the wall of your blood vessel, sticking out like a sore thumb. The immune system in your blood jumps into action and attacks the splinter, leading to inflammation and eventually an attack on the blood vessels themselves.[5,6] Lo and behold, this leads to arteriosclerosis, the heart disease that, like all others, is really an autoimmune disease.

In fact, scientists first discovered the glycocalyx when they used WGA labeled with radioactive dye to see where it went in the body and found it stuck to the glycocalyx. Wait a minute. This foreign protein was found stuck to our blood vessels, and no one bothered to pay any attention? Spoiler alert: it's actually a pretty important detail.

Remember, WGA is such a small lectin that it can get past the gut wall without leaky gut. And it doesn't stick just to blood vessels; it sticks to and actually goes through the blood-brain barrier, along with LPSs.[7] And it binds to the surface of your eyes. Dry, itchy eyes, anyone? Thank your sandwich with whole wheat bread for that. And oh, yes, WGA binds to cartilage. Hip or knee replacement, anyone?[8]

But back to heart disease and leaky gut. Sure enough, patients with atherosclerosis have reduced levels of butyrate-producing bacteria[9] to protect their gut linings. This leads to higher circulating levels of LPSs and zonulin, the protein that increases intestinal permeability.[10] But when mice were fed a high-butyrate diet for ten weeks, the atherosclerosis in their aortas was reduced by 50 percent.[11]

Further evidence of the leaky gut/leaky heart connection comes from intestinal alkaline phosphatase (IAP), which is secreted by your enterocytes. As you recall, IAP detoxifies LPSs and helps keep them from entering the bloodstream. When mice were given human IAP, they had a significant reduction in atherosclerosis.[12]

Still not convinced that atherosclerosis stems from the gut? In recent years, many studies have found bacterial DNA in atherosclerotic plaques.[13,14] Remember what contains bacterial DNA? LPSs that have slipped through a leaky gut, of course!

Earlier, I talked about the importance of the oral microbiome when it comes to producing gasotransmitters such as nitric oxide that protect against heart disease. There is now evidence that dysbiosis in the oral microbiome contributes to heart disease in other ways, too. Oral infections allow bacteria to enter the bloodstream, triggering the immune system just like the bacteria that slip through the gut wall.

Furthermore, people with oral dysbiosis are likely also to have gut dysbiosis.[15] Perhaps one too many minty gargles with that bacteria-killing mouthwash? In addition to LPSs, oral bacteria have been found in heart valves, aortic aneurysms, and arterial walls.[16] In other words, leaky mouth = leaky heart, too!

By now, you may be wondering, *What about cholesterol? I thought that was what caused heart disease.* This is what we've all been told, but it's simply not true. Because scientists found cholesterol stuck to the blood vessels in patients with heart disease, many doctors and researchers wrongly believed (and the majority of them still believe) that cholesterol had caused the plaque. But this is just like assuming that all of these various diseases cause leaky gut! *How did we get it so wrong?*

In reality, when blood vessels are inflamed, cholesterol particles come along and act as a spackling compound, covering up areas of inflammation or gaps along the glycocalyx. A study in 2023 confirmed that cholesterol has nothing to do with heart disease; heart disease stems from inflammation in your blood vessels caused by leaky gut.

Want more proof? A study in March 2023 involving more than 31,000 subjects showed that statin drugs, even if they dramatically lowered cholesterol levels, had no effect on plaque progression or on heart attacks and death. Rather, the driver was inflammation, as measured by high-sensitivity C-reactive protein (hs-CRP) and tumor necrosis factor alpha (TNF alpha) tests, two tests that I routinely do in my clinics.[17]

For a real-life example of this, let's look at a tale of two of my patients:

I have one patient who's sixty-eight years old and is the administrator at a large medical clinic here in the desert. He's not overweight, but he has extremely high cholesterol levels. When he came to see me, his low-density lipoprotein (LDL)—the so-called bad cholesterol—level was a whopping 400 mg/dl. Most doctors will tell you that your LDL should be below 100 mg/dl, so that was a significant—and scary—number.

All of that patient's previous doctors had insisted that he go on statin drugs to lower his cholesterol, but he didn't tolerate them very well. So he wanted to find another way to assess his real risk, and his cardiologist had him take a computerized tomography angiogram (CTA, as opposed to a CT calcium score, which only looks at calcium deposits in coronary arteries). A CTA shows a beautiful three-dimensional reconstruction of the coronary arteries, and—wait for it—it revealed that this patient had zero plaque in his arteries. None. His blood vessels were as smooth as a baby's butt. And since you waited for it, guess what his cardiologist's response was? Of course, his cardiologist responded to the revelation by saying that he should start taking statins in order to *avoid* developing plaques!

When that patient came to me, I explained that by looking at the blood work I had ordered, I could have predicted his angiogram findings. He had no inflammation in his blood vessels, nor did he have any markers of cholesterol oxidation. That meant that his high cholesterol levels did not put him at risk for developing heart disease. There was no inflammation on the blood vessels for the cholesterol to stick to or spackle over. Furthermore, cholesterol has to be oxidized (or activated) in order to stick to blood vessels in the first place. Oxidative stress can oxidize LDL (hence the name), making it "sticky." But that patient had been following a high-polyphenol diet and had a healthy gut, so he had no inflammation or oxidation to worry about.

Now compare this to another patient of mine, who had recently been put on a high dose of statins by his well-meaning practitioner to get his LDL very, very low. His LDL was down to a mere 37 mg/dl, but, to his chagrin, when we looked at his blood work, the particles were extremely oxidized. He had inflamed arteries, and—big surprise—severe arterial plaques. Lo and behold, he had all the markers of leaky gut, while the other one didn't. Just as in the recent study I mentioned above, all of the statin-induced LDL lowering in the world wasn't the answer to his problem; it was addressing his leaky gut that would make the difference.

How did I know? I study my patients and publish my results. For instance, in 2018, I demonstrated that dietary lectins cause coronary artery disease by prompting an autoimmune attack on blood vessels.[18] Furthermore, in nearly a thousand patients with known coronary artery disease and plaques whom I followed for a mean of nine years on the *Gut Check* program the total new cardiovascular events were 1.6 percent.[19] Contrast that to the typical 10 to 20 percent of new events at five to ten years of follow-up with maximal medical therapy (including statins).[20]

To be clear, I have nothing against using statins as a temporary intervention. They can indeed limit damage to blood vessels, but they do it in a completely different way than most patients—and doctors themselves, for that matter—realize. Statins actually block TLRs, the scanners on your immune cells that recognize invaders and tell your immune system to attack.[21] As a result, they reduce inflammation along the arteries, hopefully leaving nowhere for cholesterol to attach itself to.

But it's not just atherosclerosis. Changes in the gut microbiota and its metabolites result in increased levels of circulating LPSs that are related to myocardial infarction, heart failure, atrial fibrillation,[22] and hypertension.[23,24] Patients with every one of these conditions have dramatically reduced microbial richness and diversity in their guts and especially low levels of those all-important butyrate-producing bacteria. This is not a case of heart disease causing gut dysbiosis! For evidence of this, consider the fact that when germ-free mice were given fecal transplants from humans with hypertension, they developed hypertension.[25]

I am grateful for the fact that many of my colleagues finally seem to be getting a clear picture of what's really causing heart disease. In recent years, many studies have concluded that heart disease is the result of gut dysbiosis and decreased levels of butyrate-producing bacteria, leading to a dysfunctional gut barrier that allows LPSs, lectins, and even bacteria to leak into the bloodstream, bind to TLRs, and trigger inflammation.[26,27,28,29] Another mic drop moment! The point is, if you or a loved one has any of these "heart" conditions, the place to look for the cause and treatment is not in the chest—it's in the gut!

LEAKY GUT = LEAKY JOINTS = LEAKY METABOLISM

At the risk of sounding redundant, leaky gut is also at the root of osteoarthritis, as well as obesity, which is thought by many to be the primary risk factor behind this degenerative joint disease. In this case, invaders slip through a leaky gut and bind to the lining that is meant to protect the joints, leading to inflammation and disease. Patients with osteoarthritis in their knees have increased levels of LPSs—and the worse their condition, the more LPSs they have.[30]

Earlier, you read about how important adequate vitamin D levels are to help your gut buddies protect the gut wall. A human study looked at the gut bacteria and vitamin D level in people with and without osteoarthritis. It showed that those who were vitamin D deficient had altered microbiome compositions, with further changes in the group with vitamin D deficiency and osteoarthritis.[31] A lack of vitamin D causes changes to the microbiome and leaky gut, which leads to inflammation and osteoarthritis.

Many other human studies using probiotics and prebiotics have confirmed the gut-joint connection.[32] In one randomized double-blind placebo-controlled trial, more than 500 patients with osteoarthritis in their knees received either a probiotic supplement or a placebo every day for six months. After six months, the group that had received the probiotics saw significant improvements in their symptoms and reduced systematic inflammation.[33]

In a similar study, patients who were given prebiotic fiber saw an increase in their beneficial gut buddies, resulting in a reduction in systemic and knee joint inflammation and a preservation of cartilage.[34] Finally, a recent study showed that a combination of tamarind seed and turmeric, both polyphenols, reduced knee pain, inflammation, and cartilage degradation in patients with osteoarthritis in their knees.[35]

Even if you don't like the leaky gut theory of osteoarthritis and prefer to believe that obesity or "wear and tear" is the real root cause, you can rest assured that it is still stemming directly from the gut. In fact, there is a specific "obesity microbiome" with its own unique makeup. In particular, people with obesity have an altered ratio between the two main phyla of bacteria in our guts: Firmicutes and Bacteroidetes. These two phyla make up about 90 percent of the bacteria in our guts.

Broadly speaking, obese patients tend to have fewer Bacteroidetes and a higher-than-normal prevalence of Firmicutes. They also have reduced microbial diversity.[36] This reflects a microbiome that has shifted away from ecological balance and is no longer based on SCFA metabolism and a sharing of resources. It has transformed into a system that metabolizes simple sugars instead.

Simply put, if you have an "obesity microbiome," you have a shortage of gut buddies that work together, ferment prebiotic fiber, and produce SCFAs and an overgrowth of bad guys that feed and grow off of sugar. Of course, this type of dysbiosis leads to changes in the all-important metabolites that are a part of your gut buddies' language.[37] And the severity of the changes is correlated with the level of obesity.[38]

Back to vitamin D for a moment: When otherwise healthy women with vitamin D deficiency were given vitamin D supplements, their microbiome diversity significantly increased, and the ratio between Firmicutes and Bacteroidetes improved.[39] That helped them shift away from an "obesity microbiome" and closer to a state of homeostasis.

The "obesity microbiome" not only leads to all of the other problems associated with dysbiosis, but it also changes your hunger level and even the types of foods you crave! In addition to all of their other jobs, the metabolites created by the gut microbiome act as signaling molecules that regulate your appetite.[40] In other words, your gut buddies send messages directly to your brain, telling it when and what you need to eat—or, more accurately, what *they* need to eat. If bad guys have hijacked this messaging system, they will send the signal for you to eat more sugar because it's their fuel of choice, leading you to crave sugary foods. Yes, they not only control your hunger and appetite but control what foods you actively seek out. No wonder your appetite sometimes feels out of control.

This type of imbalance sets the stage for type 2 diabetes, which is also the result of gut dysbiosis, reduced microbial diversity, leaky gut, and chronic inflammation.[41] Scientists recently discovered that Japanese patients with type 2 diabetes even had significantly more gut bacteria in their blood—yes, in their blood!—than people without type 2 diabetes. The bacteria had translocated from the gut to the bloodstream![42] If that isn't a wake-up call, I don't know what is.

Further, metformin, the most common medication used to control signs and symptoms of type 2 diabetes, works not by lowering your blood sugar but by improving gut diversity.[43] Even berberine, which is also taken to regulate blood sugar, works by altering the microbiome, repairing the gut wall, and therefore modulating the immune system.[44] It never fails to amaze me! But again, why not go straight to the source and heal the root cause?

All of this means that if you find yourself having trouble controlling your weight, managing food cravings, or struggling with high blood sugar or even diabetes, remember that it's not your fault; it's just one more sign that you're in need of a *Gut Check*. There are many people today who eat a healthy diet and are still overweight or obese. This is because we've ignored the most important contributor to weight management: the right set of gut bugs. Ultimately, they are the ones in charge.

When we stop ignoring our gut, these issues can be reversed. When obese mice were given phlorizin, a phytonutrient present in apples, it reduced their LPS levels and mitigated the obesity-related shift in gut microbiota.[45] Further, giving obese mice prebiotic fiber restored their microbiomes, corrected the Bacteroidetes/Firmicutes ratio, reduced inflammation, and protected their knees from osteoarthritis by preventing a loss of cartilage.[46] And giving butyrate supplements to obese, prediabetic mice significantly enhanced their intestinal barrier function, reduced weight gain, and improved insulin resistance, hyperinsulinemia, and hyperglycemia.[47] I'll keep saying it until I've convinced you: when you take care of your gut buddies, they will take care of you!

LEAKY GUT = LEAKY BONES

When I tell my patients with osteopenia (reduced bone density) or osteoporosis (a more severe case of bone loss) that there is a gut issue behind these bone conditions, they generally look at me as though I've lost my mind. But it's true that even our bones are ultimately at the mercy of our gut.[48] There are immune cells in our bones just as there are in our body's tissues, with TLRs that recognize LPSs and other invaders and activate inflammatory cytokine release. It's the same old story, just sung in a slightly different key.

As with obesity, there is a specific type of dysbiosis in patients with osteoporosis.[49] And once again, the severity of disease (in this case, bone loss) is correlated with the extent of the microbiome imbalance.[50]

Germ-free mice have significantly increased bone volume compared to conventional mice,[51] and antibiotics restore bone mass in mice with osteopenia.[52,53] Don't get me wrong; I'm not saying that we should wipe out our microbiomes with antibiotics to reverse bone disease. However, these studies make clear the fact that without bacteria leaking through the gut wall and triggering the immune system, mice do not suffer from inflammation and bone loss. Of course, it's much more effective to go to the source and restore the gut instead.

We have seen many times that changes to the gut biome affect bone health and can even treat bone disease. Probiotics that reduce inflammation and improve intestinal impermeability also prevent bone loss.[54,55] The all-important SCFAs that protect the gut lining are also linked to bone health.[56] And indeed, restoring the microbiome with probiotics reduces bone loss.[57] It all really does go straight back to the gut. Again, I wonder how Hippocrates knew!

THE GUT/CANCER CONNECTION

Yes, folks, even cancer ties right back to leaky gut. Patients with colon cancer have dysbiosis,[58] reduced microbiome diversity and richness,[59] and increased intestinal permeability, leading to chronic inflammation. More and more studies point to the microbiome's role in the development of many other types of cancers, too, through its influence on the immune system.[60] With dysbiosis and leaky gut, invaders can seep through the gut wall and trigger inflammation that makes it easier for cancer cells to grow, and bad guys can take over your microbiome and enact damage that results in oncogenesis (the birth of cancer cells). One of the ways they do this is by altering the signals that tell your immune system whether or not to allow the cancer cells to reproduce and grow.[61]

Normally, your gut buddies signal to your immune system to swoop into action and kill off cancer cells before they can overgrow. But when your gut is in a state of dysbiosis, these signals are interrupted and your immune system becomes crippled, literally unable to fight off cancer. Another way to look at this is that with chronic leaky gut, your immune system is so distracted by its constant war with invaders that it doesn't have enough cops to walk the beat and spot a rogue cancer cell and kill it before it starts to grow. On top of that, the cops don't get the messages to check out local disturbances. As you recall, these messages come in the form of microbial metabolites, including SCFAs and polyamines. In patients with cancer, changes to these metabolites affect the cancer's progression and even metastasis.[62]

Even worse, when signaling molecules, or metabolites, are altered, they can become toxic and/or directly damage cellular DNA. The accumulation of DNA damage can lead to cancerous mutations.[63] For example, during tumor growth, the presence of certain inflammatory cytokines and other metabolic products can cause immune cells to reverse course and become tumor promoting instead of tumor suppressing![64] Now you have inflammation and a hampered immune system, and these allow the cancer to keep growing.

By increasing SCFA levels through probiotics administration, the signals to the immune system can be restored. This induces apoptosis (death) of cancer cells and inhibits tumor proliferation.[65] In particular, our old friend butyrate signals the expansion of T-cells to regulate the immune response and suppress both carcinogenesis and tumor growth.

But let's not forget the importance of the bacterial connection between your gut and your mitochondria. It has been noted that mitochondrial uncoupling is deregulated in many different types of cancer cells. Once again, this is thanks to the interruption of the signals being sent from your gut buddies to your mitochondria.[66] This causes cancer cells to behave in a way that is, frankly, odd.

Let's go back to our conversation about how ATP (cellular energy) is made. Back when the earth lacked oxygen, the earliest cells, which lacked mitochondria, made energy by fermenting glucose. That provided only two molecules of ATP for every molecule of glucose. Over trillions of years, the process evolved, mitochondria were put in charge of energy production, and the modern eukaryotic cell was formed. Today, our mitochondria use the method that you read about earlier to produce a whopping thirty-two molecules of ATP for every molecule of glucose. This represents a quantum leap in energy production, which ultimately allowed most plants and animals to evolve and thrive.

When oxygen is not available, cells can revert to this fermentation energy production process. Intriguingly, cancer cells do this even when they have access to oxygen. If you think about it, this makes absolutely no sense. Cancer cells want to grow rapidly, which of course requires plenty of energy. When relying on this fermentation process rather than the much more efficient electron transport system of mitochondria, they require *sixteen times* as much sugar to create the same amount of energy.

Why would cancer cells do this? Scientists previously believed that it was because cancer cells were damaged and had lost the ability to produce energy any other way than through fermentation. But now we know that this is not the case. They revert to fermentation by choice, not by necessity.

The most recent theory, and one that I like, which was originally proposed by researchers at Arizona State University, is that cancer cells do this because they no longer trust damaged mitochondria to make energy. Those mitochondria haven't been receiving signals from the microbiome to uncouple or clean up ROSs and as a result have been damaged to the point of no return. So cancer cells make energy another way, cutting the dysfunctional mitochondria out of the process.

Let me explain. Those of us who are old enough to remember the original Windows computer operating system know that if you did something wrong, the computer would balk and refuse inputs. When you shut down and restarted the computer, it would restart in "safe mode." The computer would work, but just barely, until you figured out the problem and fixed it.

In this theory of cancer, the damaged cell begins operating in "safe mode," meaning that it reverts to the primordial cell machinery that uses fermentation, not mitochondria, to make ATP. Moreover, primordial cells had no contact inhibition. Advanced cells like yours and mine stop growing and dividing when they touch their neighbor. They want their personal space! But cancer cells that run into neighboring cells just keep growing and dividing. Yup, sounds as though "safe mode" is the default system of cancer cells.[67]

Indeed, there is plenty of evidence of the connection among cancer, the gut, and mitochondria. Polyphenols, which as you know normally act to protect mitochondria from ROSs, do the exact opposite in cancer cells in order to protect you! In cancer cells, polyphenols actually increase oxidative stress and drive tumor cells into apoptosis.[68] In particular, the polyphenol resveratrol is toxic to cancer cells and has antiproliferative effects.[69,70] But of course, polyphenols can act on cancer cells only after your gut buddies have digested them for you.

This further illustrates the importance of having the right mix of gut buddies to help you fight cancer from the inside. Recently, flavonoids, one type of polyphenol, were found to help reshape gut microbiota to help fight cancer.[71] When your gut buddies digest flavonoids, they inhibit carcinogenesis.[72] Finally, butyrate and the other SCFAs produced by well-fed gut buddies directly inhibit cancerous changes, directly stop cancer cell division, and modulate the ability of the immune system to seek out and destroy cancer cells.[73]

The balance of your microbiome even determines how well you will respond to chemotherapy, which makes sense since we know that these bugs are in charge of your immune system. Your gut bugs literally modulate the efficacy or toxicity of chemo.[74,75] When mice with colon cancer were given antibiotics, the anticancer effects of chemotherapy were suppressed.[76] They did not have the right gut buddies to tell the immune system how to respond!

This is particularly complicated in cases of colon cancer, as radiation therapy can cause dysbiosis and damage epithelial cells, leading to an impaired gut barrier. This can, of course, increase inflammation and adversely affect the efficacy of other cancer treatments.[77] These are all important factors to consider when deciding how to treat cancer.

Okay, so there is a clear case for the fact that cancer stems from dysbiosis and leaky gut, but—surprise, surprise—there is even more to this story. Scientists once believed that our organs and tissues were sterile, but we now know that they actually contain microbial populations of their own, although they exist in smaller numbers than the ones in your gut. This means that the tumors within these tissues also harbor bacteria, known as the intratumor microbiome.[78] The intratumor microbiome plays a big role in tumorigenesis, disease progression, drug resistance, and even disease prognosis.[79]

Wouldn't you know it—intratumor bacterial dysbiosis is correlated with oncogenesis,[80] metastasis,[81] a suppressed immune response,[82] and drug resistance.[83] In other words, when there is an imbalance of bacteria inside cancerous tumors, they prevent your immune system from stopping the cancer cells from birthing, dividing, and dying,[84] while promoting an inflammatory environment in which cancer can thrive.[85] This is directly linked to disease prognosis.[86]

But where are the bugs in the intratumor microbiome coming from? It may not be what you think. Recent studies have shown an overlap between intratumor dysbiosis and the bacteria that are normally found in the oral microbiome.[87,88] In addition to oral cancer,[89] oral microbiome dysbiosis is associated with cancers of the esophagus, liver, stomach, breast, lung, colon, rectum, and pancreas.[90,91]

However, this doesn't leave the gut microbiome out of the picture. In some cancers, the makeup of the intratumor microbiome is markedly similar to that of the dysbiotic gut biome,[92] indicating that cancer can stem from a leaky gut *or* a leaky mouth. Furthermore—and this should come as no surprise at this point—the bacteria in the gut and the bacteria in the tumors talk to each other! These bad guys can team up to turn off the immune system and promote cancer growth.[93] This certainly seems to be the case in lung cancer, in which the lung and gut microbiomes promote cancer by inhibiting the immune system and releasing inflammatory factors.[94]

I know, it's overwhelming to consider all the ways our inner ecosystem impacts our health. But we're not done yet. As you're about to see, even your neurological and mental health is linked to the state of your gut, because at the end of the day, a leaky gut equals a leaky brain.

CHAPTER 6

LEAKY GUT = LEAKY BRAIN

Now that you better understand the importance of maintaining the sugary linings protecting our gut as well as other important parts of our bodies, it's time to look at one more sugar lining that is critical to preserving your health: the glycocalyx that lines the blood vessels going into your brain.

The blood-brain barrier (BBB) is the gateway to your brain and your central nervous system (CNS). As the name implies, it is there to prevent microorganisms and other foreign proteins from gaining access to this valuable real estate. Like your gut wall, the BBB has a layer of endothelial cells lined up next to one another. These cells regulate passage through the BBB and into the nervous system. And just like your blood vessels elsewhere, these cells are guarded by a thin layer of sugars and lipids called the glycocalyx.

The BBB also contains pericytes, cells that destroy potentially harmful microorganisms, and astrocytes, cells that recruit white blood cells into the CNS through the BBB. These two types of cells are additional lines of defense to protect your nervous system. On the other side of the BBB are microglia, the main white blood or immune cells of the CNS. The microglia are on constant alert for bad bugs or other intruders and are there to destroy any that they find—or even suspect. Think of them as the bodyguards of your rock star neurons.

The BBB functions very similarly to the gut lining with its own mucus covering. The endothelial cells can grant access to nutrients, passing them through the cell body and across the barrier. But microbes, proteins, and most drugs are normally barred from access. If microbes get through the endothelial cells and exit on the other side, they come into contact with the pericytes, astrocytes, and microglia.

With all of these lines of defense in place, the BBB should provide you with plenty of protection from all types of invaders. But just as with the gut wall, there are several ways the BBB can be breached. Some invaders, such as the bad bug *E. coli*, are able to bind to the receptors on endothelial cells and pass through these cells and across the BBB. But in general, the glycocalyx must be damaged in order for this to happen. And—surprise, surprise—that nasty lectin WGA that is present in whole grains is perfectly capable of completing this damaging process.[1] In fact, WGA can facilitate crossing of the BBB![2] And don't forget about those LPSs. They, too, open the gates across the BBB. So if you have LPSs and WGA on the loose, the "force fields" guarding your brain are literally down, leaving your normally protected brain open and vulnerable to attack.

A permeable BBB also leaves your entire nervous system more vulnerable. Indeed, in the case of multiple sclerosis (MS), an altered BBB allows various immune cells to infiltrate the CNS, leading to inflammation and injury to the myelin sheath that protects the nerves. No surprise, patients with MS tend to have altered microbiomes, with a reduction in butyrate-producing bacteria.[3]

Another way that invaders infiltrate the BBB is via the so-called Trojan horse method. In this case, a pathogenic invader infects a white blood cell that has been recruited into the CNS via the BBB and piggybacks across. They sure are clever! It seems that for each line of defense, invaders have developed a new strategy to break through the BBB. In case you are wondering (you are, aren't you?), there are blood tests that can measure the breakdown of the BBB and test you for "leaky brain."

Of course, it's worth noting that none of this would be an issue if we had a robust ecosystem and intact gut wall to begin with. Let's be clear about this: the only reason these invaders have any opportunity to breach the BBB is because they have already breached the gut or oral cavity linings. This may seem obvious, but we must never forget where all of the problems are really stemming from. And that's sadly what I see again and again in my patients' blood work: if they have a leaky gut or mouth, most of them also have a "leaky brain."

What happens when one of these sneaky invaders slips through the BBB? Essentially the same thing that occurs when the gut lining is breached: inflammation and resulting disease. Brain fog, forgetfulness, full-blown neurodegenerative diseases such as Alzheimer's and Parkinson's, and mental health conditions such as anxiety and depression are all by-products of inflammation in the brain.[4,5]

Over the past few years, there has been a worrisome increase in the number of people, including young people, who are struggling with mental health. Many are quick to blame the stress of the covid-19 pandemic (and it certainly didn't help, as strong evidence suggests that covid infection causes leaky gut and alters the microbiome[6]), but I believe the present mental health crisis, though heightened by the stresses of the pandemic, has a deeply physiological basis. When you have dysbiosis and leaky gut, no part of your health or well-being is safe, and that certainly includes your mood, your brain function, and your mental health.

THE BRAIN-GUT AXIS

So the BBB is pretty important. How can we protect it? By having the right gut buddies and avoiding WGA, of course! We now know that the microbes in our guts regulate the BBB in a number of different ways. One is via the vagus nerve, the main nerve of the parasympathetic nervous system, which plays an important role in mood, immune response, digestion, and heart rate.

Just like your gut biome, the vagus nerve is essential in maintaining homeostasis. It runs all the way from the gut to the brain, and your gut buddies use it like a telephone wire along which to send signals upstairs about what's going on down in the gut. These signaling molecules include various metabolites, neurotransmitters, gasotransmitters, and hormones, many of which you read about earlier. They tell your immune system whether or not to attack intruders[7] and instruct your brain how to think, feel, and act.[8]

The signals travel in both directions, but roughly 90 percent of them are sent up from the gut to the brain instead of the other way around.[9] Read that again: your gut sends *90 percent* of the signals to your brain, with a measly 10 percent going from your brain to your gut. Remind me again who's really in charge?

I think it's fair to conclude that a happy gut equals a happy brain—in addition to a happy, healthy body, of course. When you have the right mix of gut buddies, they send messages up the vagus nerve saying that everything is going well: the immune system can stand down, your nervous system can relax. They'll also produce postbiotics to help maintain the BBB. The SCFA butyrate protects the BBB just as it does with the gut wall. Germ-free mice have increased BBB permeability compared to mice with healthy gut biomes. But get this: when the germ-free mice receive either a fecal transplant from healthy mice *or* supplements of SCFA-producing bacteria, the integrity of their BBB is restored.[10]

With leaky gut and/or dysbiosis, on the other hand, invaders can begin seeping through the BBB while your gut sends the message along the vagus nerve that you are under attack. In this case, instead of battening down the hatches and reinforcing its line of defense, the BBB becomes more and more permeable, allowing more invaders in, while the brain and the entire nervous system become more and more inflamed. This is the environment in which neurodegenerative diseases can thrive, along with depression, anxiety, and other mood disorders.

You already know about some of the signaling molecules that make up your gut buddies' language. These, such as SCFAs, travel from the gut to the brain and directly affect the nervous system. Butyrate not only protects the BBB but also enhances cholinergic neurons, which play an important role in brain function and release an essential neurotransmitter, acetylcholine.[11] Butyrate can also cross the BBB and activate the vagus nerve and the hypothalamus,[12] and it has been shown to have antidepressant effects in animals with depression and mania.[13,14]

However, there are additional postbiotics I haven't yet discussed that are particularly important when it comes to brain function and mental health. These are neurotransmitters, which send signals between neurons. That's right; your gut buddies play a large role in determining the levels of neurotransmitters in your brain,[15] which directly impacts your cognitive function.[16] For example, germ-free mice have alterations in the fecal and serum levels of many neurotransmitters.[17,18] As you probably now suspect, antibiotic administration creates the same types of changes.[19,20]

Since some neurotransmitters cannot cross the BBB, your gut buddies send their precursors to the brain instead. These are the raw materials needed to produce those neurotransmitters. In the brain, your neurons and glial cells pair these precursors with other enzymes to synthesize neurotransmitters. Interestingly, your gut buddies can produce the same enzymes and use them to produce neurotransmitters in the gut, too.

Wait a minute; why do your gut buddies bother producing neurotransmitters when they cannot cross the BBB? To act locally on the millions of neurons in the gut, of course! Though many researchers have taken to calling the gut the "second brain," I would argue that it actually comes first. In fact, there are more neurons in your gut than there are in your entire spinal cord. Talk about a "gut feeling"! The neurons in the gut are connected to the ones in your brain, and the neurotransmitters made in the gut also act as fast signals to the brain via the vagus nerve.[21,22,23] This is another way that your gut buddies alter brain chemistry and influence your mood and behavior.

Even patients with ADHD have an altered microbiome that creates excess ROSs and altered apoptosis. Interestingly, patients with ADHD have an increase in a specific type of bacteria that produces a precursor to dopamine,[24] the neurotransmitter that regulates reward anticipation, a process that does not function normally in these patients.

Studies on zebrafish, which have a genetic structure similar to that of humans, reveal that the microbiome affects social behavior by stimulating the remodeling of microglial cells in the forebrain during early brain development.[25] These microglial cells also regulate neurons and synapses, the junctions between neurons.[26] The creation of the networks in the brain is essential to developing normal social behaviors.

Like most immune cells, the microglia that are in charge of brain development and synapse regulation respond to signals from your gut buddies.[27] This means that if you don't have the right mix of gut buddies, your brain won't get the proper signals about how to develop and form the networks. This directly impacts brain development and whether or not you are able to exhibit normal social behaviors.[28] Sadly, if the microglial cells don't get the right messages—or worse, if they get messages that the brain is under attack—they will break the connections between neurons in a process called pruning and even eat them whole.[29]

It's a similar story when it comes to cognition. Signals from the gut affect neural circuits and connectivity, exerting control over cognitive functions. Simply put, a more balanced microbiome leads to improvements in cognition, brain function, and emotional well-being.[30]

Before we go further—and yes, there is much more to this story—let's take a closer look at the neurotransmitters that your gut buddies help produce as another part of their complex language.

Glutamate

Glutamate is the most abundant excitatory neurotransmitter, meaning that it prompts neurons to share information with one another. It is responsible for sending signals between nerve cells and is involved in neuroplasticity, learning, and memory.[31] Neurons and astrocytes can make glutamate using metabolites produced by your gut buddies as precursors.[32] Cells in the intestinal tract can also produce glutamate and use it to send rapid signals to the brain via the vagus nerve.[33]

Glutamate has a direct impact on your mental health. Altered levels of glutamate in plasma, serum, cerebrospinal fluid, and brain tissue are associated with mood changes, psychotic disorders, and even the risk of suicide.[34,35,36] Now, don't panic. Once you have the right gut buddies in place, they will help maintain healthy levels of glutamate in your brain and your gut.

Gamma-Aminobutyric Acid (GABA)

Gamma-aminobutyric acid (GABA) is an inhibitory neurotransmitter, meaning that it blocks messages from being sent from one neuron to another. It counterbalances the action of glutamate, and having a low level of GABA is linked to depression and mood disorders.[37]

In the brain, GABAergic neurons produce an enzyme that converts glutamate into GABA.[38] Therefore, healthy levels of glutamate are necessary to have healthy levels of GABA. Other metabolites produced by gut buddies, including the SCFA acetate, are also a part of the GABA production process. These can cross the BBB into the hypothalamus so that GABA can be produced in the brain.[39]

Acetylcholine

Acetylcholine sends excitatory signals between neurons.[40] It is produced by several bacteria,[41,42] but it cannot cross the BBB and luckily can also be synthesized by neurons.[43] However, even the acetylcholine made in the brain is dependent on your gut buddies. As I mentioned earlier, butyrate enhances the neurons that produce it! Without the right gut buddies producing butyrate, those neurons cannot produce adequate acetylcholine.

A case in point: patients with Alzheimer's disease often have altered levels of acetylcholine in their brains.[44] And in mouse models of Alzheimer's disease, butyrate not only ameliorated symptoms and improved memory but also reversed memory deficits by regulating the metabolism of astrocytes, one type of immune, neuron-supporting cells of the brain.[45,46]

Dopamine

Earlier, you read about how the single-celled organism toxoplasma hijacks dopamine production in rodents to make them less risk averse. Well, your gut buddies play a big role in your dopamine levels, too. Dopamine and dopamine receptors are widely distributed in the intestinal tract, and more than half of the dopamine in the body is produced by the gut.[47]

Patients with depression have reduced transmission and uptake of dopamine.[48] Plus, your gut's relationship with dopamine is bidirectional: your gut buddies produce dopamine, and your dopamine levels affect gastric secretion, motility, and mucosal blood flow.[49]

Serotonin

Serotonin is considered to be the key neurotransmitter to the gut-brain axis, as it affects the birth of new neurons that express both dopamine and GABA.[50] Abnormal expression and function of serotonin in the brain are associated with depression and anxiety disorders.[51] Proper levels of tryptophan, an amino acid that is a precursor of serotonin, are also critical to mental health. When patients with depression who had been in remission experienced a temporary reduction in their tryptophan level, their symptoms returned.[52]

Approximately 90 percent of the serotonin in the body is produced outside the brain, primarily in the intestinal epithelium. Spore-forming bacteria promote this process by producing a metabolite that signals an increase in gene expression along the tryptophan pathway.[53,54] Some of your gut buddies can even use tryptophan to make serotonin directly![55]

As you read earlier, the glyphosate in the weed killer Roundup targets those gut buddies' health. And since it is routinely sprayed on most wheat, corn, soy, canola, and oats in North America (yes, even non-GMO crops), many researchers believe that all of the healthy "whole grain goodness" is delivering more and more Roundup right to our gut.[56]

DEPRESSION, ANXIETY, AND THE GUT

With your gut buddies playing such a big role in the production of neurotransmitters, it's no surprise that there is a direct link between gut dysbiosis and depression and anxiety. Over the past few years, we have begun to see evidence that the microbiome can cause depression and anxiety via the signals they send along the gut-brain axis that create—you guessed it—inflammation.

Patients with major depressive disorder (MDD) have a significantly different microbiome makeup from that of healthy individuals, with reductions in gut diversity and richness and changes in the ratio of Firmicutes to Bacteroidetes.[57] Specifically, patients with MDD have enriched pro-inflammatory bacteria and depleted anti-inflammatory, butyrate-producing bacteria.[58,59] Patients with generalized anxiety disorder (GAD) also have significant decreases in microbial richness and diversity, with reductions in SCFA-producing bacteria. Yet patients in remission for GAD no longer have these changes![60]

We have seen the direct relationship between depression and the gut time and time again. When germ-free mice were given fecal transplants from patients with MDD, they began exhibiting behaviors associated with depression. They also developed disturbances in their microbial genes and metabolites![61] Their gut buddies were no longer sending the right signals to their brains, and the result was that they became depressed.

In 2022, a groundbreaking study looked at the microbiomes of more than 1,000 patients with depression. They found alterations in thirteen bacteria that were associated with depression. Those bacteria are known to be involved in the synthesis of glutamate, butyrate, serotonin, and GABA. In addition, patients with depression have depleted levels of butyrate-producing bacteria.[62] And in patients with severe mental illness, increases in dysbiosis, zonulin level, LPSs, and inflammation are correlated with the severity of the disease.[63]

Additional studies have shown that both anxiety and depression are linked to dysbiosis and specifically to a microbiome that secretes excess LPSs. No surprise, this leads to systematic inflammation with major ramifications in the brain, resulting in neuroinflammation.[64] These gut-related brain disorders are not limited to anxiety and depression. When patients with eating disorders, including anorexia nervosa, bulimia nervosa, and binge-eating disorders, took probiotic supplements, they experienced a reduction in symptoms.[65]

We've also recently learned that a decrease in hydrogen sulfide (H_2S), one of the gasotransmitters produced by your gut buddies, is associated with depression.[66] As you recall, H_2S regulates synaptic plasticity, the strengthening of the connections between neurons. Without the right gut buddies producing H_2S, the synapses can become weak, the neurons lose their ability to communicate with one another, and this can cause depression.

If you want even more evidence that the gut plays an important role in mental health, look no further than this: the class of drugs known as selective serotonin reuptake inhibitors (SSRIs), which are the predominant psychopharmaceutical treatments for depression, work by making changes to the gut and even have direct antibacterial effects. Several different strains of bacteria metabolize SSRIs. This leads directly to changes in the makeup of the microbiome.[67]

A recent study compared the gut makeup of a group of patients with MDD compared to healthy controls. Before treatment, the patients with MDD had significantly reduced microbiome diversity and richness compared to the healthy controls. After the group with MDD received treatment with an SSRI (escitalopram, which is sold under the brand name Lexapro) and their symptoms had improved, they were tested again. This time, the microbiomes of the treated patients and the controls were not significantly different.[68]

Furthermore, SSRI treatment can lead to some of the other benefits of a healthy microbiome, including weight loss. When another SSRI (fluoxetine, better known as Prozac) was given to healthy male mice, they saw significant, time-dependent changes to the microbiome along with a reduction in body mass.[69]

Amazingly, traditional Chinese medicine (TCM), which has been around for thousands of years, also treats depression through the gut. TCM uses bioactive compounds that regulate the gut microbiome to improve levels of neurotransmitters, SCFAs, and cytokines. These changes exert antidepressant-like effects.[70] Once again, I am astounded by the ancient wisdom that we sorely need to tap back into.

NEURODEGENERATIVE DISEASES AND THE GUT

With the gut exerting such thorough control over the brain, it should come as no surprise that neurodegenerative diseases, including Alzheimer's and Parkinson's, start in the gut, too. The mechanisms of these diseases might seem a bit more complex, but they ultimately also stem from gut dysbiosis, leaky gut, and the resulting chronic inflammation and autoimmune activation.

Long before most of us were aware of the gut/Parkinson's disease connection, it was widely known that gastrointestinal dysfunction was a common early symptom of Parkinson's disease (PD).[71] Patients with PD often suffer from constipation and have increased intestinal inflammation. In fact, these GI symptoms often appear before patients develop any of the motor symptoms that we associate with PD.[72]

Now we know that these GI symptoms stem from gut dysbiosis, which, of course, causes changes to patients' microbial metabolites.[73] In particular, patients with PD have a reduced abundance of short-chain fatty acid (SCFA)–producing bacteria and, as a result, lower concentrations of SCFAs to do their many jobs, most importantly protecting the gut lining and the BBB, all while signaling to uncouple mitochondria, keeping neurons healthy.[74,75]

Patients with PD also have reduced expression of tight-junction proteins, worsening leaky gut. This allows invaders into the bloodstream, where they trigger inflammation and eventually cross the BBB to cause mischief in the nervous system. Sure enough, patients with Parkinson's have higher-than-normal concentrations of LPSs and inflammatory markers, and increases in these levels are correlated with the severity of their disease.[76,77]

So there are leaky gut, dysbiosis, leaky brain, and widespread inflammation. This certainly sets the scene for a neurological condition to develop, but it is also where it gets a bit more complex. In addition to all of these factors, neurodegenerative disorders, including both PD and Alzheimer's disease (AD), also feature the accumulation and aggregation of misfolded proteins. Simply put, protein folding is when the proteins in your body transform from a chainlike structure into a three-dimensional structure. This is how proteins become biologically functional.

For many reasons, including certain genetic mutations, a protein can be misfolded into an incorrect shape that is dysfunctional.[78] This is a fairly common occurrence. Mistakes happen, and there are cellular "quality control" systems in place that look out for faulty proteins. Normally, when quality control finds a misfolded protein, it degrades and disposes of it, and all is well.[79] But when too many proteins are misfolded, the quality control system is overwhelmed and can't keep up. The misfolded proteins start to build up, leading to organelle dysfunction and cell death.[80]

A hallmark of PD is a buildup of misfolded alpha-synuclein proteins that accumulate in between neurons. These clusters of misfolded proteins are called Lewy bodies.[81] What does all of this have to do with the gut? It turns out that patients with PD also have Lewy bodies in between the neurons in their guts![82] Plus, it is gut dysbiosis that is causing all of the protein misfolding.[83] When bad guys overgrow, they can secrete bacterial amyloids that cause oxidative stress and induce the aggregation of proteins.[84,85]

Bacteria produce amyloid proteins to help them bind to one another, forming biofilms.[86] This helps them resist being killed off by the immune system.[87] When bad guys or any bacterial population that is overgrowing sticks together this way, it becomes even more dangerous.

One type of bacteria, *P. aeruginosa*, is overabundant in patients with both PD and AD. This bug is able to damage the gut wall directly,[88] release its own aggregating proteins,[89] and trigger the release of the amyloid proteins that are associated with these degenerative diseases.[90] There is another type of LPS-producing bacteria called *Desulfovibrio* that does much the same thing and has been found in the guts of patients with PD.[91] Besides accumulating between neurons in the gut, the resulting protein aggregates travel from the gut to the brain via the vagus nerve.[92]

Complicating the picture even more, although bacterial amyloids are different from the amyloids produced in the brain, they have a similar structure.[93] And when the immune system gets used to reacting to bacterial amyloids, it stays on high alert and begins attacking the amyloids that are produced in the brain, too.[94] This is another case of molecular mimicry leading to an inflammatory autoimmune response, this time in the brain![95]

If you're still not convinced that PD starts in the gut, consider this. Researchers in Sweden analyzed the PD risk of hundreds of thousands of people, including those who had received a procedure called a vagotomy, which involves cutting one or more branches of the vagus nerve. There are two types of vagotomy: a truncal vagotomy, which fully transects the nerve, and a selective vagotomy, which cuts only the part of the nerve going to the stomach. The data showed that the patients who had received a truncal vagotomy were *40 percent* less likely to develop PD than patients who had not received a vagotomy or had received a selective vagotomy. There was no significant difference in the risk levels between the two latter groups.[96]

This makes perfect sense, since the bacteria-derived amyloids and resulting protein aggregates travel to the brain from the gut, not the other way around.[97,98] With a full truncal vagotomy, there is no highway for them to travel along and no way for them to reach the brain. They are forced to remain in the gut and damage the neurons there, causing GI issues rather than neurological disease.

The development of AD is a similar story. AD is characterized by the accumulation of misfolded amyloid beta plaques between neurons. These are tangled together by tau proteins. At face value, this sounds exactly like PD but with a different protein—and it is—but we have recently discovered that amyloid beta is actually an antimicrobial peptide that is part of the immune response.[99]

In a state of homeostasis, amyloid beta helps fight against bacterial infection, but when there are leaky gut and leaky brain, it becomes problematic. Patients with AD have high levels of LPSs in their bloodstream and even in their brains.[100] When TLRs on the microglia (the immune cells of the brain) are constantly activated by LPSs, they begin to respond similarly to amyloid beta.[101] Worse yet, the microglia actively eat away at the connections between the nerves, destroying the brain and memory.[102] It's yet another case of molecular mimicry,[103,104] leading to neuroinflammation and neurodegeneration! When researchers manipulated the gut microbiome of mice with tau protein–related brain injuries, they observed a strong reduction in brain inflammation and brain damage.[105]

Further, patients with AD are missing important lipids in the brain called plasmalogens, a decrease in which is associated with chronic inflammation.[106] Interestingly, inulin, a sugar found in chicory, one of my favorite greens, can be converted into certain plasmalogens. I'll give you one guess as to who else loves chicory. Your gut buddies, of course! They convert it into plasmalogens to help protect your brain.[107] In order to benefit, though, you need chicory in your diet and the right mix of gut buddies to digest it for you.

GENETICS MATTER—BUT NOT IN THE WAYS WE THINK

You may be wondering: What about the role of genetics in developing a neurodegenerative disease? It's a good question, and to be honest, the answer really surprised me. We know that the apolipoprotein E4 (APOE 4) genotype is the greatest genetic risk factor for developing AD. Most people carry the APOE 3/3 genes, but there is a mutation that causes people to carry the APOE 4 gene. About one in four people carries one copy of APOE 4. This doubles the risk of developing AD. It's much rarer to carry two copies of APOE 4, but some people do, and this raises the risk of developing AD more than twelvefold.[108]

The APOE gene helps metabolize fat, and with the gene mutation, cholesterol can't be moved into and out of cells properly. You can think of it as a train carrying passengers to a station. When the train (APOE) arrives at the station (which in this case is a cell), the passengers (cholesterol) get off the train and go into the station. Normally, if it is the passengers' final destination, they stay there. Everyone else gets back onto the train to go elsewhere. With this mutation, all of the passengers are dropped off at the station and stay there. They can't leave. The passengers build up, and the station gets overcrowded. In non-metaphor terms, cholesterol builds up inside the cell. So although dietary cholesterol isn't a problem for most people, those with the APOE 4 gene should limit their production of cholesterol.

But it doesn't stop there. There are, after all, other important lipids in the body. People with this mutation also don't have enough docosahexaenoic acid (DHA) in their brain, even though their blood level is normal.[109] This is important because DHA makes up half of the membrane content of our neurons![110] This happens because people with APOE 4 don't produce enough apolipoproteins, which transport lipids. So there is no train to carry the DHA to the brain. People with AD, whether they have the genetic mutation or not, benefit from taking fish oil (which contains DHA) and apolipoproteins (the later being present in Krill Oil).[111]

By now you're probably wondering what any of this has to do with the gut biome. Well, until recently, I didn't think it did! For years, I've been saying that it's not your genes that control your fate; it's your gut buddies. To a large extent, this is true. But as I was doing the research for this book, it was incredibly eye opening for me to discover that there is in fact a genetic component to the type of microbiome that lives inside us.[112,113] In other words, our genes play a role in determining our microbiome profiles.

This genetic component is relatively small in the grand scheme of things. Regardless of your genes, you have a great deal of control over your microbiome, as it is highly sensitive to dietary and environmental changes. But the genetic component does play a role and should not be ignored, particularly when it comes to mutations such as APOE 4. It turns out that the APOE 4 genotype is associated with a specific microbiome profile that includes significant reductions in amino acid and SCFA production.[114] Son of a gun! I can't believe that I missed this connection until now.

Though the APOE 4 mutation can seem scary, this should also give you a lot of hope if you or someone you care about carries the gene. First of all, as scientists learn more about this, I envision an incredible future of personalized medicine. Second, understanding the gut/APOE connection enables us to intervene with supplementation and alterations to the microbiome, such as seeding it with butyrate-producing bacteria. In a mouse model, manipulating the microbiome led to significant decreases in inflammation, tau protein pathology, and resulting damage from AD.[115]

And don't forget, although we now know that genetics impact the microbiome, the microbiome in turn impacts gene expression. It is truly a two-way street.[116] As I keep saying, it takes two. And we will tackle it from all directions in the *Gut Check* program.

CHAPTER 7

LEAKY GUT = LEAKY HORMONES

Another complex and fundamental way in which your inner ecosystem controls your health and well-being is through your endocrine system, a delicate balance of hormones that circulate through your bloodstream, sending messages that regulate your organs and many of your bodily functions. The importance of having the right balance of hormones cannot be overstated, and, as I hope you have now come to expect, your gut buddies play many different essential roles in producing, balancing, and metabolizing your hormones. Yes, this is another layer of the language that your gut buddies use to communicate and exert their control and influence over the rest of your body.

Certain gut buddies metabolize hormones and then either reactivate and recirculate them into the bloodstream or expel them as waste. This directly affects the circulating levels of many of your hormones. Some gut buddies produce the precursors that you need to produce hormones. Others metabolize various hormones to produce those hormone precursors. Each has a role to play, and in a stable, healthy inner ecosystem, they all come together perfectly.

With leaky gut and dysbiosis, however, microbiome-driven changes to circulating hormone levels can have massive consequences for your health. We see this most dramatically with the three main sex hormones—testosterone, progesterone, and estrogen—as well as the stress hormone cortisol. In fact, so intertwined are your hormones and your microbiome that there is a whole new field of study called microbial endocrinology.

Is Your Stress Coming from Your Gut?

Cortisol is a steroid hormone produced by your adrenal glands that is released into your bloodstream when you are stressed, and a lot of people are confused about the role it plays in the body. Many of my patients come to me convinced that they're suffering from "adrenal fatigue," a condition in which the adrenal glands are taxed following prolonged periods of stress. Many of them also believe that their weight gain is a result of high cortisol levels. But as I've written before, I measure all of my patients' fasting a.m. cortisol levels, and it's very rare for me to find elevated or diminished levels.

What research does show, however, is a direct correlation between levels of the gut buddy *Ruminococcus* and three other chemicals: N-acetylaspartate (NAA), which is responsible for brain energy production; serotonin; and cortisol.[1] Astonishingly, the symptoms of "adrenal fatigue" are more likely the result of a gut issue rather than of overworked adrenal glands.

Meanwhile, the stress you are feeling is probably not the result of elevated cortisol levels, either. A recent human study from 2022 showed that perceived stress was not correlated with blood levels of stress hormones such as cortisol. What was it correlated with? Microbial dysbiosis![2] Stress alters the microbiome, which can cause leaky gut, which can trigger inflammation and neuroinflammation. The reverse scenario is just as potent: intestinal dysbiosis coupled with leaky gut causes inflammation, neuroinflammation, and a drop in NAA production, causing feelings of stress.3 The good news is that with a Gut Check, your ecosystem and your moods can return to an even keel.

THE GUT/HORMONE SYMBIOSIS

The preceding chapter should have you convinced that your hormones and your microbiome have a bidirectional relationship. Your gut buddies affect your hormone levels, and your hormone levels can also exert direct changes to your microbiome. For instance, scientists have known for decades that the hormone progesterone promotes the growth of the *Bacteroides* genus[4] of bacteria, and we now know that changes to the microbiome also affect progesterone levels.[5] In women, a low progesterone level can lead to infertility, irregular menstruation, anxiety, and depression, whereas a high progesterone level can cause anxiety and depression, along with weight gain and fatigue. We rely on a balanced inner ecosystem to keep these levels in check.

Meanwhile, when men's testosterone level goes down, it leads to changes in their microbiome.[6] At the same time, certain bacteria promote testosterone metabolism and reabsorption in the colon. This is an important part of regulating testosterone levels.[7]

Some gut buddies can even produce testosterone by metabolizing another type of steroid hormone called glucocorticoids.[8] Men with a high testosterone level can experience high blood pressure, headaches, and excessive body hair, but today it is far more common for men to have a low level of testosterone. This can lead to infertility, sexual dysfunction, muscle weakness, fatigue, and depression.

Women need testosterone, too. If their testosterone level is too low, they might experience the same symptoms as men do. When their testosterone level is too high, symptoms such as excess body hair, balding, infertility, and voice deepening can occur. A high testosterone level can also contribute to and/or be a consequence of polycystic ovary syndrome (PCOS). And since progesterone is a precursor of testosterone, it's essential to have adequate levels of progesterone, too.

But the microbiome has perhaps the most direct and impactful relationship with the hormone estrogen. In men, the symptoms of too much or too little estrogen are similar to each other. They include sexual dysfunction, infertility, and depression. Another common symptom of a high estrogen level in men is enlarged breast tissue, technically called gynecomastia but better known as "man boobs," something that I see all too often in my patients these days. And wouldn't you know it, gynecomastia correlates with the serum level of phthalate (a major endocrine disrupter found in plastics and chicken) in boys.[9]

In women, the estrogen level naturally decreases after menopause. In premenopausal women, low estrogen can lead to brittle bones, irregular menstruation, weight gain, dry skin, and thinning hair. When the estrogen level is too high, women can experience headaches, worsening premenstrual symptoms, weight gain, and fatigue.

The estrogen level directly impacts the gut, too. As you may know, when their estrogen level naturally rises during pregnancy, women experience rapid changes to their microbiomes.[10] Perhaps not surprisingly, premenopausal women have a distinctly different microbiome makeup from men's, but postmenopausal women have a microbiome that is more similar to men's.[11] There is plenty of evidence that this is caused by the decreased estrogen level in postmenopausal women.[12]

To be sure, estrogen itself has positive effects on the microbiome. It can help reduce gut permeability and therefore LPS circulation.[13] In mice, estrogen makes the gut wall more resistant to injury, and this, of course, leads to a reduced inflammatory response.[14] As women go through menopause and their estrogen level decreases, they often experience greater intestinal permeability, which (no surprise) leads to an increase in inflammation.[15]

Since the relationship between the microbiome and hormones is bidirectional, these changes to the microbiome during menopause can further alter women's hormone levels. If you're starting off with an intact gut wall and a well-balanced microbiome, your gut buddies will help manage this process for you. Otherwise, it can become a vicious cycle of dysbiosis leading to imbalanced hormone levels, leading to greater dysbiosis, leading to even more imbalanced hormone levels! This leaves you more susceptible to obesity, metabolic syndrome, endometriosis, and even certain types of cancer.[16]

Thankfully, you can reverse this cycle by encouraging a stable ecosystem, leading to balanced hormone levels, which will further support the stability of the microbiome. But for now, because it is so important, let's focus on how the gut impacts estrogen levels, which in turn directly affect the gut.

THE ESTROBOLOME

We tend to think of estrogen as a "woman's hormone," and it does play an important role in women, particularly during their childbearing years. However, men have (and need) estrogen, too. In women, estrogen's main job is to tell the cells to store fat in case of an upcoming pregnancy. This is why excess estrogen in men leads to "man boobs" and often a pregnant-looking belly. It also regulates the menstrual cycle and affects the reproductive tract, urinary tract, heart, blood vessels, bones, breasts, skin, hair, mucous membranes, pelvic muscles, and, last but definitely not least, the brain. In men, estrogen also plays an important role in sexual function and the production of sperm.

It's clear that having the right amount of estrogen is important, but most people don't realize that the microbiome directly regulates the levels of estrogen in the body. We have known since the 1980s that the use of antibiotics leads to a reduction in estrogen level,[17] and in recent years we have begun to more fully understand how this occurs. Now we refer to the estrobolome, the group of recently discovered bacteria in the microbiome whose products are capable of metabolizing and modulating estrogen.[18]

The estrobolome works primarily by producing an enzyme called glucuronidase. Normally, once it has done its job, estrogen is broken down in the liver and released into bile, which then enters the intestine to be expelled as waste. Glucuronidase halts that process and essentially puts the broken-down estrogen back together, making it once again biologically active. This reactivated estrogen is then sent back into the bloodstream.[19]

Glucuronidase is not necessarily bad. We need some of it, but having the right balance is of utmost importance. And of course, glucuronidase activity is dependent on the composition of your microbiome.[20] A stable ecosystem equals balanced glucuronidase and therefore a balanced estrogen level. But if your microbiome is in a state of dysbiosis and you have excess glucuronidase-producing bacteria, it will reactivate and send too much estrogen back into circulation.[21] The opposite is also true: too little glucuronidase equals not enough estrogen.[22]

Estrobolome dysbiosis, which leads to altered levels of glucuronidase and therefore changes to estrogen levels, is in fact behind many diseases. For example, increased estrogen leads to altered mitochondrial and immune system function, which is common in women with endometriosis.[23] Lo and behold, increased estrogen exposure can stimulate the development and progression of endometriosis.[24]

However, perhaps the greatest danger of estrobolome dysbiosis is an increased risk of developing cancer. Of course, any type of dysbiosis increases the risk of developing cancer for all of the reasons I previously discussed—namely, by inducing chronic inflammation and disrupting the immune response.[25] In addition to those changes, too much estrogen dramatically increases the risk of developing a specific type of cancer called estrogen receptor–positive cancer. We see these most often in the breasts and ovaries.

Some (but not all) cancer cells contain estrogen receptors. When estrogen attaches to these receptors, the cancer cells grow. Basically, these particular cancer cells feed on estrogen. Women who have a high concentration of circulating estrogen—particularly after menopause, when the estrogen level normally decreases—are more susceptible to this type of cancer.[26]

Not only does activating estrogen receptors allow cancer cells to grow, it damages the mitochondria in those cells.[27] This promotes cancer growth even more. So estrogen receptor activation can be a deadly one-two punch. Since the estrobolome plays such a big role in regulating the circulating level of the estrogen that activates these receptors, the health of the estrobolome is a large determinant of estrogen receptor–positive cancer risk.[28]

Women who are at risk of developing estrogen receptor–positive cancer are often given the drug tamoxifen, which acts on estrogen receptors to reduce cancer growth. When mice were given tamoxifen, they developed elevated levels of *Lactobacillus*, a helpful, anti-inflammatory gut buddy. Remember, the estrogen levels impact the microbiome, and vice versa. The researchers then gave *Lactobacillus* to a separate group of mice, which went on to experience a decrease in tumor formation.[29] This highlights the intricate connections between our gut buddies and our hormones and how their connections can impact cancer risk. Regardless of the type of treatment that a patient with cancer pursues, this connection should not be overlooked.

One more time (well, in reality I'll probably say it a few more times): when you treat your gut buddies well, they will take care of you. But the opposite is also true. Though your gut buddies normally activate polyphenols to create beneficial signaling molecules, they can also use polyphenols to synthesize estrogen-like compounds that activate estrogen receptors.[30] This means that if you don't have the right mix of gut buddies, all of the polyphenols that you're eating might end up working against you, increasing your risk of developing certain cancers instead of protecting you!

THE BREAST AND OVARIAN MICROBIOMES

As in all other cancers, breast and ovarian cancer tissues have their own microbiomes. Cancerous breast tissue has a distinct microbiome compared to healthy breast tissue, with increases in certain bacteria and decreases in others. Interestingly, there are consistent changes in cancerous breast tissue when compared to adjacent noncancerous tissue from the same patient and to breast tissue from other, healthy women.[31]

In 2021, the largest study ever to look at healthy versus cancerous breast microbiomes found some fascinating connections between changes to the breast microbiome in cancer patients and alterations in the immune system.[32] In general, decreased microbial diversity in the breast microbiome led to immune system dysregulation. More specifically, certain gut buddies, such as *Propionibacterium* and *Streptococcus*, were depleted in cancerous breast tumors. These bacteria in particular are related to the activation of T-cells, the type of white blood cells that are on the first line of defense against cancer cell detection and eradication. In other words, breast microbiome dysbiosis with a reduced abundance of these specific gut buddies leads to a decrease in T-cell activation. This allows cancer cells to grow.

Furthermore, many of the bacteria that are underrepresented in breast cancer tissue produce important metabolites. These include cadaverine, a somewhat ominous-sounding compound that nonetheless inhibits breast cancer invasion,[33] and our good friend butyrate, which we already know is anti-inflammatory and also has antitumor effects.[34] In general, patients with breast cancer have reduced levels of SCFA-producing bacteria in their gut.[35]

Remember, SCFAs are not just anti-inflammatory; they are some of the most important signals that your gut buddies use to communicate with your immune system. It is clear that this line of communication is essential to activating the immune system against cancer while also avoiding the chronic inflammation that can aid in cancer growth. Moreover, butyrate and the other SCFAs are histone deacetylase (HDAC) inhibitors, anticancer agents that directly prevent cancerous cell changes and halt cancer cell division.

Ovarian cancer, too, is characterized by a specific intratumor dysbiosis. In this case, it extends beyond the tumor tissue to the upper and lower female genital tract, pelvis, peritoneum, and intestines. Notably, this particular brand of dysbiosis includes a higher-than-normal population of gram-negative bacteria, leading to an increase in LPSs and resulting in inflammation.[36] This LPS-induced inflammation is a major driver of ovarian cancer growth.[37,38]

In addition, microbes associated with both pelvic inflammatory disease[39] and other sexually transmitted diseases[40] trigger inflammation and cause changes to the ovarian microbiome. This can impact the induction and progression of ovarian cancer, too.

Meanwhile, changes to the intratumor microbiome and its metabolites directly impact the growth and death of these cancer cells. When ovarian cancer cells are treated with SCFAs, they are driven to apoptosis,[41] and when antibiotics are administered to ovarian cancer cells, they block cancer cell proliferation and reduce the proportions of ovarian cancer stem cells.[42] Perhaps most impressive, when mice with ovarian cancer were given supplements of *Akkermansia*, the friendly gut buddy that helps protect the gut lining, it significantly suppressed ovarian cancer growth by activating the immune system against cancer cells.[43]

It's worth repeating, so I'll say it yet again: all diseases begin in the gut. When your inner ecosystem is in a state of homeostasis and maintains a healthy line of communication with your immune system and your mitochondria, your hormone levels and the microbiomes within your tissues will also remain stable. They will recruit the immune system only when necessary (for instance, to kill cancer cells), keeping inflammation low and your body healthy. This is the state that we are meant to live in and that we are going to re-create, but first let's look at one more thing that is currently keeping us from this beautiful inner balance.

XENOESTROGENS

Complicating matters, there are endocrine disrupters—the chemicals I discussed earlier that mimic or interfere with the body's hormones—in many of our everyday products, including plastic bottles, the liners of metal food cans, detergents, flame retardants, toys, cosmetics, pesticides, and even foods.

Endocrine disrupters don't impact only the exposed individual; they can create epigenetic changes that alter gene expression and lead to transgenerational consequences.[44] Here's one example that's extremely compelling, albeit somewhat extreme. In 1976, a chemical factory explosion in Italy exposed nearby residents to an endocrine disrupter called dioxin. The exposed women experienced decreased fertility—as did their daughters who had been exposed in utero![45]

One major category of endocrine disrupters is xenoestrogens, which can bind to estrogen receptors and change the expression of estrogen-related genes. As you can imagine, when xenoestrogens mimic estrogen in the body and bind to estrogen receptors, it can dramatically increase the risk of developing estrogen receptor–positive cancer. This means that even if your microbiome and estrobolome are in perfect balance, you still might be at risk because of high levels of circulating xenoestrogens. Of course, if you're starting out with leaky gut and dysbiosis, your risk is even higher.

In fact, xenoestrogens are more damaging than regular estrogen. Unlike naturally occurring estrogens, which bind to receptors and then detach, xenoestrogens bind to receptors and stay there, continually activating those receptors without giving them a break. Of course, continually activating the estrogen receptors on cancer cells enables them to continue to grow and grow.

It's incredibly important to get xenoestrogens out of your life and your body. Let's take a look at a few of the different types of xenoestrogens and how you can rid yourself of them to minimize your risk.

Parabens

These are preservatives that are used in many items, including beer, sauces, sodas, and many personal care and cosmetics products. Parabens have been detected in human tissues and bodily fluids and have been found in high quantities in breast tissues in patients with breast cancer.[46] In animals, parabens can increase breast cancer cell proliferation and tumor size.[47,48] Paraben levels also correlate strongly with prostate cancer and its aggressiveness in men.[49] Read ingredient labels, folks, and avoid anything with a term that ends in "paraben," such as methylparaben and propylparaben.

Phthalates

These chemicals are found in hundreds of products, from plastic wraps to plastic containers, plastic bags, chicken meat, flooring materials, and personal care products. They can bind to estrogen receptors, increasing breast cancer risk,[50,51] depress sperm count and testosterone levels in men, and are correlated with diminished penis size when boys are exposed in utero.[52] What's alarming is that many of us now have measurable levels of phthalates in our bodies. It's no surprise that phthalates also cause diabetes.[53]

To avoid phthalates, choose fragrance-free products and avoid plastics of all kinds as much as you can! Visit butcher counters where the meat, chicken, and fish are still wrapped in "butcher paper." Store your food in glass containers. (Some good news, ziplock bags are phthalate free.) And this just in: your fast-food meal, especially burritos, hamburgers, and chicken nuggets, is likely full of phthalates.[54] I can just hear it now: "Would you like to supersize your breast or prostate cancer with your order today?"

Nonylphenols (NPs)

These toxic substances are used in various industrial processes and are found in consumer laundry detergents, personal hygiene products, automotive products, latex paints, and lawn care products. They exhibit estrogen-like activity on estrogen receptor–positive breast cancer cells[55,56] and have been found in human breast milk, blood, and urine. Again, check those ingredient labels. These things are hiding in many places.

Bisphenol A (BPA)

Believe it or not, BPA was first used as a pharmaceutical estrogen back in the 1930s. More recently, it has been used to manufacture certain plastics. It is found in food containers, water bottles, and other protective coatings.[57] It should come as no surprise that something that was developed to mimic estrogen in the body actually mimics estrogen in the body! BPA disrupts estrogen receptor activity by mimicking, enhancing, or inhibiting the activity of natural estrogens.[58] Its use is strongly correlated with an increased risk of developing breast, prostate, and uterine cancers.[59]

In the United States, BPA use has been forbidden in some products but is allowed in low levels in others. To protect yourself, avoid canned foods and foods stored in plastic as much as possible, and opt for glass or stainless steel instead. As I said earlier, its replacements, including BPS and BPF, are no safer.

Dichlorodiphenyltrichloroethane (DDT)

DDT is a pesticide that was used for many years across agricultural fields and homes. It is an insect neurotoxin that kills mosquitoes and other insects that carry malaria, typhus, and other diseases. Though it is banned in the United States and some other countries, it is still widely used in places where malaria is endemic, particularly in India and southern Africa.[60]

DDT accumulates in adipose tissue and interferes with estrogen receptors.[61] This increases breast and ovarian cancer risk.[62] In men, DDT exposure leads to a decrease in semen volume, concentration, and motility.[63,64]

After DDT was banned, methoxychlor (DMDT) was synthesized as an alternative to protect pets, crops, and livestock from pests such as mosquitoes, cockroaches, and other insects. It inhibits estrogen binding and is also associated with an increase in ovarian cancer risk.[65] After many years of use, DMDT is now banned in both the United States and Europe.

Quaternary Ammonium (Chlormequat)

Chlormequat is an herbicide/pesticide that is sprayed on oats and other grains to keep their stalks short so they won't break or bend in the wind. (I'm not making this up!) It is an endocrine disrupter, as well as being associated with birth defects.[66] It is so dangerous that it is not approved for spraying on food crops in the United States, but in 2018 the Environmental Protection Agency (EPA) allowed traces of it to be present in US oats, wheat, and barley. In 2020, the Trump administration further increased the levels permitted in US oats.

Recently, the Environmental Working Group (EWG) tested eleven oat products, including several Quaker Oats products, Cheerios, and several granolas, and found that all of them had dangerous levels of this chemical. My advice for many other reasons and now this one is to avoid products containing oats.[67]

• • •

By now, I hope I've convinced you of a few things. One, your gut buddies are more manipulative and intelligent than you ever previously imagined. Two, they exert their control over every aspect of your health and well-being in incredibly complex, multifaceted ways—right here, right now, as you are reading these words. And three, their actions will determine whether you fall prey to disease or live a long, happy, healthy life.

The good news is that your relationship with your gut buddies is meant to be symbiotic. You scratch their back, and they'll scratch yours (so to speak). So let's move on and take a look at how exactly you can start taking better care of them so that they, in turn, will take good care of you.

CHAPTER 8

CIGARETTES, MEAT, AND CHEESE

The Secrets of Longevity Are Not What You Think

When I wrote *The Longevity Paradox: How to Die Young at a Ripe Old Age*, it was clear to me that many of us had misinterpreted which factors truly contribute to a long and healthy life, particularly regarding our assumptions about some of the most famously long-lived cultures around the world. But as the research and my own perspective have evolved, I now realize that I didn't quite understand the whole picture then. And by now it will be no surprise that I have come to believe that the biggest differentiator between people who live long and healthy lives and those who don't is the health of the ecosystems in their guts.

Case in point: a recent study looked at the microbiomes of centenarians (aged 99 to 104 years) and semi-supercentenarians (aged 105 to 109 years) and compared them to the microbiomes of younger adults.[1] Of course, the first two groups had dramatically different microbiomes compared to the last. Of particular interest was the fact that the centenarians and semi-supercentenarians had increased levels of bacteria that were able to metabolize xenobiotics (chemical substances).

What does this mean? People who live to a ripe old age have a microbiome that is adapted to protect them from all of the plastics and endocrine disrupters you read about earlier. Their microbiomes are so robust that they can basically handle anything that's thrown at them. In fact, their gut buddies actually eat these foreign compounds in the gut before they can get into the bloodstream and exert any damage. (As an example, there are even oil spill–eating bacteria that can render those toxins harmless!) Unfortunately, the rest of us can't handle these xenobiotics. We are so badly affected by them because we've obliterated the ecosystems inside us. Perhaps that's why most of us don't make it to age 109.

Before you raise your hand to protest that these folks just have great genes, let me be the first to say that you're right. But it's not their human genes that make the difference. As you know, the vast majority of the genetic material inside you is bacterial, not human. So, yes, they do have great genes. They just happen to be great bacterial genes, not human!

This is yet more proof that your gut buddies have a vested interest in keeping you around. You are their home, and they are very adaptable little beings. They have the ability to rapidly mutate, gain new genetic information, and evolve in order to protect you. But they cannot do this if they are not present in large, diverse populations, with healthy communication systems between themselves and the rest of your body.

I'm sure you're already asking yourself how these superagers managed to nurture such robust inner ecosystems. Well, keep reading, because it may not be what you think.

THE MYTH OF THE BLUE ZONES

Let's back up for a minute and look at some of the mistaken assumptions we've made about what factors create longevity and, just as important, how. About fifteen years ago, the journalist Dan Buettner wrote a book called *The Blue Zones: Lessons for Living Longer from the People Who've Lived the Longest*, which was based on reporting that he had done with some of his colleagues looking at the villages around the world that (allegedly) had the longest-lived people. Those villages were labeled Blue Zones.

Why blue? A researcher (who will go unnamed here) suggested (but did not prove) that certain places had a higher-than-normal number of centenarians. At a meeting in Montpellier, France, he circled those locations on a world map with a blue felt-tip pen. These "Blue Zones" included the Ogliastra region on the Italian island of Sardinia; Okinawa, Japan; Loma Linda, California (where I once lived and worked as a professor at Loma Linda University); the Nicoya Peninsula of Costa Rica; and the Greek island of Ikaria. Not included in Buettner's list were a few other locations whose residents are famously long lived, including the Kitavans in Papua New Guinea and the residents of Acciaroli, a small town south of Naples, Italy.

I have spoken with Dan Buettner, and I have a lot of respect for his work. The problem with the Blue Zones theory of longevity, however, is twofold. One is that the data themselves are imperfect and incomplete. It also reflects Buettner and his colleagues' biases.[2] In fact, when you look closely, much of the data seem to have been selected to fit a predetermined narrative.

Paul Simon said, "A man hears what he wants to hear and disregards the rest." Buettner is an ardent mostly vegan vegetarian and I believe that he unintentionally disregarded some important truths because he was overly focused on the less impressive evidence that confirmed what he'd hoped to see. In truth, the overwhelming majority of people in the Blue Zones are not vegan, vegetarian, or even close. And the types of animal products they do consume are an important and long-overlooked factor contributing to their long and healthy lives. More on that in a moment.

But speaking of those long lives, the accuracy of the age records that were used to measure these people's life spans is unclear. In fact, a recent review of the Blue Zones data surmised that the majority of them may include errors.[3] It turns out that less than 15 percent of the supercentenarians (people older than 100) included in the studies had a valid birth or death certificate. Well, that doesn't bode particularly well for accuracy, does it?

Okinawa, Japan, is a good example. Okinawa appears to have the highest number of people between the ages of ninety and ninety-nine in Japan, but it also has the fewest senior citizens per capita in the country, as well as one of the highest poverty and crime rates. What is going on here?

Let's take a look at history for the answer. In 1945, the US bombing and invasion of Okinawa during World War II destroyed many birth and death records. That meant that members of the US military (most of whom didn't speak much Japanese) had to fill in replacement documents based on reports by individuals—an imperfect system at best. Is it possible that those military service people guessed at birthdates or just filled in the same or similar years repeatedly, which makes it now appear that there is a disproportionate number of people born within the same decade?

Furthermore, it may not be a coincidence that in addition to Okinawa, the islands of Sardinia and Ikaria are poverty stricken. At first glance, this doesn't make much sense. Normally, people living in richer areas tend to live the longest because they have access to better health care and nutrition. However, it's important to note that these three alleged Blue Zones were included because of extreme outliers instead of their high average longevity rate. In other words, they had an impressive number of extremely old people (supercentenarians) driving up the average, but the majority of people didn't have a greater-than-average life span.

Still, that doesn't explain why these socioeconomically challenged areas would have such a high concentration of supercentenarians. That said, we do know that in some underresourced regions, deaths are underreported. Sometimes, when a person who is receiving a pension passes away, his or her family, who is dependent on the person's income, doesn't report the death. So it is possible that some deaths were never reported or recorded.

The two remaining Blue Zones, Loma Linda and the Nicoya Peninsula, were included for their high average longevity instead of the existence of outliers. However, their inclusion on the list is just as questionable as the others. For instance, Loma Linda is a small town of just 23,000 people with an average life span of eighty-six for women and eighty-three for men. In reality, this hardly ranks high. Many other places around the world have higher average life spans, including Hong Kong, Singapore, and even some other neighborhoods in the United States![4,5] In fact, the highest life expectancy in the world is in a tiny country in the Pyrenees Mountains between Spain and France called Andorra, and the second highest is in Monaco. And wait till you see what those guys eat! Is it possible that Buettner included Loma Linda because it's the one Blue Zone where many people are vegans and vegetarians? Meanwhile, the Nicoya Peninsula in Costa Rica is not really one cohesive region; it was created by proponents of the Blue Zones theory by circling areas on a map[6] with that blue pen—almost as if it had been gerrymandered.

None of this is to say that Buettner or his colleagues intentionally deceived the public, or that we can't learn a thing or two from these regions. At the end of the day, it is true that many people in these areas do seem to live relatively long, healthy lives compared to people in other parts of the world. But this leads me to problem number two with the Blue Zones theory, which is the fact that so many people have come to the wrong conclusions about *why* these populations are allegedly so long lived. For example, many people saw that two of these areas were on Mediterranean islands and jumped to the conclusion that the secret of a long life lies in the Mediterranean diet, particularly its purportedly heavy reliance on whole grains.

Like so many other things, we have gotten this completely backward. We have assumed that consuming lots of whole grains protects these people from other negative aspects of their lifestyles, such as heavy smoking. (Ninety-nine percent of the men in Ikaria smoke!)[7] In reality, it is the exact opposite. Though the Mediterranean diet does have many benefits, those grains are actually a negative component of their diet.[8] In other words, those folks live long, healthy lives *despite* eating so many grains, not because of it. Plus, it's not as if they completely avoid suffering from eating those grains. In fact, Italians overall have high rates of arthritis,[9] and Sardinians in particular have a high proportion of autoimmune diseases.[10] There's your whole grain goodness for you!

It's time to take a closer look at what *really* helps long-lived people thrive into their golden years.

THE BLUE ZONES PARADOX: LONGEVITY AND CHEESE

You're likely familiar with the so-called French paradox, the fact that the French eat a lot of saturated fat, primarily in cheese, but have low rates of heart disease and obesity. The same thing could be said of many of the Blue Zones. Their residents eat a lot of dairy products, but, importantly, not made from cow milk. In fact, three of the Blue Zones (Ikaria, Sardinia, and Nicoya) are home to many sheep herders, and the people in those locations eat tremendous amounts of sheep milk products, including milk, cheese, and yogurt.

When my friend Dr. Mark Hyman came back from a trip to Sardinia, he said to me, "You're not going to believe this: these guys eat yogurt and cheese every day!" Well, not only do I believe it, but it's obvious to me that this is exactly what helps those people live long, healthy lives.

Consider this: It's not the entire island of Sardinia that is a "Blue Zone." The only part of the island that has great longevity is the mountainous region, which also happens to be the area where the people herd sheep.[11] Similarly, most of the people in Costa Rica eat a lot of beans and grains, but not in Nicoya, whose residents herd sheep, as well. And Ikaria—you guessed it, the people are sheep and goat herders. And it's the population of Seventh-Day Adventists in Loma Linda who actually live a long time and drive up the area's average life span. Seventh-Day Adventists live an average of ten years longer than the average American, and their diet is made up of a full *50 percent* dairy products! I should know. When I arrived there, I berated the dietitians in the hospital, telling them that all the yogurt and cheese they were serving was killing my heart surgery patients! Boy, was I wrong.

What the heck is going on with all this cheese? Why aren't all those people suffering from heart disease from all that saturated fat and cholesterol? Well, I've already addressed the myths behind cholesterol. Further, these dairy products are indeed high in saturated fat, but it is a special type of saturated fat: medium-chain triglycerides (MCTs). Sheep, goat, and water buffalo milk contains 30 percent MCTs! As I mentioned earlier, MCTs are powerful mitochondrial uncouplers.

People in France, too, eat plenty of sheep cheeses, including Manchego, pecorino, and feta, as well as A2 casein cow dairy products. Is it possible that the French remain slim and free of heart disease *because* of all the cheese they eat and not *in spite of* it? Is that what's really at the heart (no pun intended) of the French paradox? I am tempted to say yes. And so do recent human studies.

The story doesn't end with MCTs, although that alone might be a good enough reason to add sheep, goat, and water buffalo dairy products to your diet. The milk from these animals (and products made from them) also contains an additional uncoupling compound called milk fat globule membranes (MFGMs), which surround the fats in these milks and make them soluble. As mitochondrial uncouplers, MFGMs aid in weight loss and insulin resistance.[12,13] And dare I say that even Harvard nutritionists have trouble getting around the fact that consumption of high-fat ice cream reduces diabetes risk.[14] So why don't you hear more about this? Because it goes against the conventional and party-line wisdom and certainly ruins a good "Blue Zone" narrative. In light of this, and with what's to come, maybe we should start calling them "White Zones."

Milk products in and of themselves are clearly important for mitochondrial uncoupling, but things really start to get interesting when cheeses are aged—aka fermented. Aged cheeses contain the important information from dead bacteria that you read about earlier, as well as postbiotics such as polyamines, which, you will recall, are compounds that help modulate the immune response and protect the gut wall. Polyamines are also mitochondrial uncouplers.[15] Perhaps this is why several studies, including a new one out of Sweden, have shown that although consumption of nonfermented milk and butter is associated with higher all-cause mortality, consumption of fermented milk and cheese is associated with lower all-cause mortality.[16,17] I'm tempted to drop the mic yet again!

As I said earlier, another benefit of polyamines is that they increase the activity of intestinal alkaline phosphatase (IAP), which, you read earlier, breaks down lipopolysaccharides (LPSs) so they can no longer cause you harm. It turns out that aged cheeses give you a double whammy of IAP. Various components in raw milk and dairy products *and* the fermentation products in cheese all stimulate the gut to produce more IAP.

Scratch that; molded cheeses actually provide a triple whammy of IAP! Cheeses such as Roquefort contain fungi that produce IAP themselves.[18] Molded cheeses also include additional metabolites that inhibit cholesterol biosynthesis and modulate bacterial growth. As such, they protect cardiovascular health.[19] And when diabetic mice were given a specific gut buddy commonly found in yogurt and cheese, they experienced a drop in their fasting glucose levels.[20]

Unfortunately, many people do have allergies or sensitivities to cow milk and other dairy products, but even this can be alleviated by fermentation. Many studies have shown that the fermentation process actually degrades the allergens in milk, including casein![21] This is akin to the way fermenting lectin-heavy foods reduces lectin levels and the way the bacteria in superagers' guts eat xenobiotics. It appears that during the fermentation process, bacteria eat the milk allergens for you so that their friends in your gut aren't harmed by them! And they eat the lactose in milk, rendering it lactose free, as well. Is there anything they didn't think of?

Unfortunately, this does not mean that you should run out and start eating a ton of conventional dairy products made from cow milk. If you've read my other books, you already know that most cow milk products in the United States come from a breed of cows that produces milk with a highly inflammatory protein called A1 beta-casein. Plus, cow milk does not contain those precious mitochondrial uncouplers, MCTs.

All milk contains MFGMs, but it's only in the milk's fat. Could it be that in our mistaken and now disproven belief in low-fat dairy products, the skim milk we've been giving our children, which contains no MFGMs and is mostly sugar, has deprived generations of a slimming, diabetes-fighting substance that long-lived farmers somehow knew was healthy? I highly suggest taking a note from long-lived farmers and consuming plenty of full-fat sheep and goat milk yogurt and cheese, instead of the skim and fat-free varieties.[22]

BEANS AND RICE ARE NOT SO NICE

As I said, one of the biggest misconceptions about the "Blue Zones" is that the people in these communities eat a lot of whole grains and beans. The Costa Ricans in Nicoya don't eat beans and grains as people throughout the rest of the country do. This is one thing that sets them apart and protects them from lectins, including WGA. When mice were fed beans, their gut biome diversity actually decreased, along with their numbers of a specific gut buddy called *Oscillospira*[23] that is associated with being lean.[24] Simply put, beans don't feed the gut buddies associated with being lean.

Perhaps even more surprising, the Okinawans don't eat the most popular grain in Japan: rice. This is a huge differentiator and clearly one of the reasons for their long life spans. But wait a minute; why don't Okinawans eat rice? It's a staple throughout Japan and a huge part of the culture. Lo and behold, it's simply because of Okinawa's unique climate. Okinawa consists of a string of subtropical islands that regularly experience severe tropical storms. Rice simply does not grow well in that climate.

In the 1600s, Okinawa imported purple sweet potatoes from China and found that they were hardy and could survive the tropical storms.[25] So the humble sweet potato became a staple of the Okinawan diet.[26] In fact, it is Okinawans' main source of fuel, making up more than 85 percent of the calories they typically consume in a day. That's a lot of sweet potatoes!

Like other root vegetables, including taro root and yams, the sweet potato is not a regular carbohydrate; it is a form of resistant starch. The name *resistant starch* refers to the fact that it "resists" quick digestion and is not immediately converted into glucose. Instead, it makes it past your small intestine and into your large intestine mostly intact.

Once they arrive in your large intestine, resistant starches are a wonderful source of nutrition, not for you, but for—you guessed it—your gut buddies. When they eat resistant starch, your gut buddies multiply and produce large amounts of the SCFAs acetate, propionate, and butyrate. Resistant starches therefore increase your gut buddy population, enhance digestion and nutrient absorption,[27] and foster the growth of the gut buddies that nurture the all-important mucus layer lining your gut. In addition, because they are not quickly digested into glucose as other starches are, resistant starches don't raise your blood sugar or insulin levels. But they do keep you and your gut buddies feeling nice and full.[28,29]

In addition to resistant starch, purple (or blue) sweet potatoes contain compounds that act as powerful antioxidants.[30] They are also rich in particularly beneficial polyphenols, such as 4,5-di-O-caffeoylquinic acid,[31] that protect the mitochondria better than do many other polyphenols, including the ones in grape skin, red cabbage, elderberry, purple corn, and ascorbic acid.[32] Polyphenols from sweet potatoes have also been shown to suppress the growth of human cancer cells.[33] But remember, all the polyphenols in the world can't help you if you don't have the right gut buddies to activate them.[34] Luckily for the Okinawans, sweet potatoes provide both resistant starch and polyphenols.

It's not just the Okinawans, though. The Kitavans of Papua New Guinea eat huge amounts of taro root, another form of resistant starch that is rich in antioxidants and polyphenols. In rats, taro inhibits the proliferation of colon cancer cells in a dose-dependent manner.[35] The Kitavans also eat a ton of coconuts, which contain those wonderful MCTs.

It's clear that these people protect and nurture their inner ecosystems through the types of carbohydrates they eat—resistant starches—and the types they don't—grains and beans. This is just one more piece of the longevity puzzle.

GET YOUR FIBER FROM MEAT

It is true that the people in the Blue Zones generally don't eat huge amounts of red meat, but they do eat some. Yet they suffer from very little coronary artery disease. This is yet another Blue Zone and French paradox. But although Buettner and others have been hyperfocused on the small amount of red meat consumed in these areas, they missed something important about the red meat the people do eat: much of it is fermented.

On the Greek island of Ikaria, it's common for families to hunt for wild goats and then consume the meat throughout the year. How do they make the meat last all year? By fermenting it! They do the same thing with pigs, which they slaughter during festivals early in the year and then ferment so the meat will last until the next year.[36]

People in these cultures also make and eat a lot of cured, aka fermented, sausages, which, of course, are popular throughout Europe. In addition to all the other benefits of fermentation, this process produces many beneficial compounds, including cadaverine, which, you read earlier, helps prevent cancer,[37] and those important polyamines spermine, spermidine, and putrescine,[38] which (among other things) help the cells in the gut wall proliferate.

I've always wondered why charcuterie is such a big part of traditional European meals. It's yet another paradox: Europeans eat all of these cured meats and cheeses that are full of saturated fats, yet they have low rates of coronary artery disease. Well, now we know why! And those long-lived Andorrans? They eat sausages and sheep cheeses daily and have the highest life expectancy in the world.

Even some of the nonfermented forms of meat that these long-lived people eat also nourish their gut buddies. Here's a term that you may not have heard before: *animal fiber*. This is the food for gut buddies that is lurking in some animal products. Unfortunately, it's the stuff that most of us discard without eating, such as the ligaments, tendons, bones, and cartilage in meat and fish. These actually act similarly to resistant starch in your gut: they resist digestion in the small intestine and are gobbled up by your gut buddies in the large intestine. Salmon skin and tendons and the ligaments between chicken bones are also good sources of resistant starch.[39]

As a little tease of something I'll talk much more about later, this is only one reason that fish and chicken are the healthiest forms of animal protein for both you and your gut buddies.

EAT YOUR POLYPHENOLS, AND FEED THEM TO YOUR FOOD

This should come as no surprise, but one thing that these long-lived people have in common is that they consume tremendous amounts of polyphenols. In the case of Okinawans, these are not just any polyphenols;[40] they're flavonoids, which you read earlier are particularly helpful in fighting cancer. Because of Okinawa's geography and subtropical climate, the people there grow unique plants that are rarely seen elsewhere.[41,42] Plus, the soil in Okinawa is made of ancient coral, which gives plants that are grown on the island exceptionally high levels of flavonoids.[43] Indeed, the more polyphenols Okinawans consume, the longer they live.[44]

These people also consume lots of polyphenol-rich spices. People in the rest of Japan call them the spice eaters. For example, turmeric, which contains the polyphenol curcumin, is traditionally used in soups, curries, and tea.[45]

Meanwhile, in Sardinia, people eat plenty of high-polyphenol fruits and veggies, but one particularly powerful source of their polyphenols is wine. We already know that moderate red wine consumption is good for our gut buddies because of its fermented polyphenols, but traditional Sardinian Cannonau wine appears to be even better. Cannonau are black grapes that grow in Sardinia and have extremely high levels of polyphenols.[46] Wines made from Cannonau grapes have two to three times the amount of polyphenols as do other wines!

The other "Blue Zones" prioritize polyphenols, too. Being mostly vegetarian or pescatarian, the Seventh-Day Adventists in Loma Linda eat tons and tons of leafy greens, and the Nicoyans grow and eat lots of polyphenol-rich tropical fruit.[47] On the Greek island of Ikaria, people grow wild greens called horta in almost every garden and cook it in olive oil, giving them a double whammy of polyphenols.[48]

Another hidden source of polyphenols in Ikaria is traditional Greek coffee. Boiled Greek coffee, called Elliniko kafe, has more polyphenols and less caffeine than other types of coffee, and most Ikarians drink it several times a day. They also eat a lot of polyphenol-rich herbs, including rosemary and oregano. Finally, the Ikarians eat large amounts of a common weed that grows in sidewalk cracks called purslane (a cousin of moss rose or portulaca), which not only is rich in soluble fiber but contains high amounts of alpha-linolenic acid (ALA). In fact, it was that observation that prompted the famous Lyon Diet Heart Study, which

I wrote about in my previous books. Heart patients who were supplemented with ALA had dramatic reductions in new coronary events compared to those following an American Heart Association low-fat diet.[49]

Back to rosemary: it is also especially popular in Acciaroli, a fishing village on the Italian coast. Locals eat the extraordinarily pungent wild local rosemary that grows there every day, including it in almost everything they cook. In addition to being full of polyphenols, it was recently learned that it contains ursolic acid, which helps regulate the gut biome![50]

In Acciaroli, people don't just eat this wonderful rosemary; they feed it to the animals that they raise and ultimately consume. The same is true of the other long-lived farming communities on the list: they feed their animals tons of local greens and herbs. As I often say, *You are what you eat, and you are what the thing you're eating ate*. The small amount of animal protein that long-lived people eat is basically a delivery device for more polyphenols. When they ferment the meat to make it last, thereby fermenting its polyphenols, all the better!

Melatonin: It's Not Just for Sleep

You're likely familiar with the hormone melatonin as a sleep inducer, and, yes, it can help you sleep, but likely not for the reasons that you think. As I said earlier, melatonin is one of only two antioxidants that act to protect mitochondria. It is made in the brain from the amino acid tryptophan (which, you will recall, is also a precursor of many important neurotransmitters). However, melatonin is also present in many different plant compounds, including leaves, stems, roots, fruits, and seeds.

What the heck is melatonin doing in those plants? They don't need to sleep. It turns out that melatonin is an antioxidant for the plant's equivalent of mitochondria, the chloroplasts. Plants produce it to protect themselves and their mitochondrial equivalents, and when we eat those plants, we reap the same benefits.

But what about sleep? Well, remember the old credo: association does not mean causation. For ages, scientists assumed that melatonin induced sleep because they saw that its level goes up at night. However, nighttime is also when mitochondrial repair happens. At night, we generate ketones, which tell our mitochondria to repair and make more of themselves. Of course, it's a good idea to have melatonin handy at that time to help clean and spruce up the mitochondria!

Wouldn't you know it, many of these long-lived people eat tons of melatonin-rich foods, such as red wine, olive oil, mushrooms, nuts, and spices. This is one more overlooked reason the Mediterranean diet is so effective.

BASK IN THE SUN

One thing that I absolutely agree with Dan Buettner about is the impact that vitamin D has on these long-lived people. It's no coincidence that four out of five of the Blue Zones are areas that receive optimal vitamin D levels through sun exposure year-round. Vitamin D is closely linked to longevity.[51] Okinawa is the one exception, as its winters tend to be dark and cloudy.

Why is vitamin D so important? As you read earlier, you need adequate vitamin D in order for the stem cells hiding in the crypts of your microvilli to proliferate and replace dead cells along your gut wall. Vitamin D impacts the gut in many other ways, too. A higher vitamin D level leads to increased gut diversity and more butyrate-producing bacteria,[52] corrects the Bacteroidetes/Firmicutes ratio, and increases the abundance of some particularly helpful gut buddies, such as *Akkermansia* and *Bifidobacterium*.[53]

So take a note from these long-lived people and get out in the sunshine, or supplement with vitamin D_3, the active form of the vitamin. If you're worried about developing skin cancer, you can always do both. People who take vitamin D supplements have fewer cases of melanoma than those who don't.[54]

If, like most people, you don't live in a climate that provides access to sunshine 365 days a year, you can get your vitamin D from another source that is common in long-lived cultures. You guessed it: dairy products, particularly sheep milk dairy products. According to the British Nutrition Foundation, sheep milk has more than four times the amount of vitamin D as cow milk![55] This is just one more way that these long-lived people take good care of their gut buddies and are rewarded with long, healthy lives.

VITAMINS AND CIGARETTES DO MIX

Warning: there's controversial content ahead. It's worth noting that many people in the Blue Zones smoke quite a bit, yet they do not suffer from any of the common consequences of smoking. In fact, smoking seems to be something that gives them more longevity, not less. For instance, in Sardinia, the men and not the women typically smoke, and the men are the ones who live longer, healthier lives. The same goes for Ikaria, Acciaroli, and Costa Rica.

How can this be? It turns out that nicotine is a powerful mitochondrial uncoupler![56] Perhaps this is why smokers are famously thin. Think back to the French paradox. Is this one additional reason why French women stay slim despite eating so much saturated fat?

Don't get me wrong; smoking is terrible for your health, but it's the delivery device that's the problem. Nicotine itself appears to have many health benefits. A study of 30,000 British doctors showed that smoking reduced the incidence of Parkinson's disease by 30 percent,[57] and other studies have shown a correlation between smoking and a reduced risk of developing dementia.[58]

You may be wondering: Even if nicotine is an uncoupler, how do these people avoid the health consequences of smoking? The answer lies in the rest of their diet. Tobacco smoking causes mischief because it induces oxidative stress. In your body, vitamin C works to protect your cells from that oxidative stress. But if you smoke, it's easy for your store of vitamin C to get used up. Then you are left unable to counter the effects of oxidative stress from smoking.

Pardon the brief detour here, but humans are one of the few animals that can't make vitamin C. Five enzymes are needed to convert glucose into vitamin C, and we have only four. Animals that make vitamin C have zero coronary artery disease. It's that protective.

Bill Sardi, PhD, who studied vitamin C extensively, showed in rats how important it is to have adequate levels of vitamin C. When he silenced the fifth enzyme needed to produce vitamin C (the one that humans are lacking), the rats lived half as long as other rats that did not have that gene silenced. But when he added vitamin C to the drinking water of the rats with the silenced gene, they returned to their normal life span.[59] Sardi believed that if humans were to have vitamin C in their diet continuously, they could live for up to 250 years.

Why can't our bodies make this all-important vitamin? What happened to the fifth enzyme? One theory is that the mutation that silenced the last enzyme was protective. If you get plenty of exogenous vitamin C in your diet, as humans in the jungle did, the glucose that would go into making it could be put to better use. Evolution tends to favor efficiency. The problem, of course, is that now most of us don't get nearly enough vitamin C from our diet to counter the oxidative stress in our bodies, whether or not we smoke. But if we were to get enough vitamin C, it's possible that we would not only be a lot healthier overall but also be immune to the negative effects of smoking.

I believe that this is exactly the case with the smokers in the "Blue Zones." Of course, a high level of vitamin C also increases longevity in the people who don't smoke. For example, bitter melon, which is remarkably high in vitamin C and is used as a medicinal herb, is a staple of Okinawan cuisine.[60] Many of the other fruits and vegetables that are consumed in the Blue Zones also provide lots of vitamin C.

Of particular note, olive oil, a staple of the Mediterranean diet, is a wonderful source of vitamin C. In fact, a compound in olive oil called hydroxytyrosol (HT) can increase your vitamin C levels by twofold![61] Add this to the long list of reasons I'm such a big fan of olive oil.

Of course, nicotine is highly addictive, as any tobacco company executive knows. But did you know that the plants in the nightshade family, which include tobacco, tomatoes, potatoes, eggplant, and peppers, all contain nicotine? Did you ever wonder how a family of New World plants, introduced into world trade only five hundred years ago, could become some of the most widely eaten (or smoked) foods throughout the globe? Are we eating these foods for the longevity boost or the nicotine "hit," or have we just become addicted to them? I report; you decide. Yes, nicotine use is a slippery slope, but it's a slope that I think we should consider revisiting instead of shunning. Of course, if you do smoke—or even if you don't—make sure that you are getting plenty of vitamin C!

• • •

At the end of the day, there continue to be many differing opinions and theories about which factors actually lead to longevity and which ones don't. But more and more people are gaining clarity about the fact that our gut buddies control our aging. Indeed, the folks in the Blue Zones take good care of their gut buddies and mitochondria by consuming plenty of mitochondrial uncouplers, melatonin, polyphenols, and fermented foods, as well as by getting lots of vitamin C and D. But there is another thing that these long-lived people all do for their gut buddies that we haven't yet discussed, and that leads me to the next chapter: they all eat according to the natural cycle of the seasons.

CHAPTER 9

TO EVERYTHING THERE IS A SEASON

Another way that we have damaged our gut buddies over the past several decades is by feeding them far too much and far too often—not to mention feeding them the wrong things! Our gut buddies certainly need nourishment in order to grow and thrive and take care of us. But they are not meant to be constantly and consistently eating throughout the day or even throughout the year. Yet this is what we've been forcing them to do. And now we are paying the price.

When we constantly feed ourselves, and hence our gut buddies, our mitochondria lose their metabolic flexibility, their ability to generate ATP using different types of fuel. As I often say, our mitochondria are meant to operate like hybrid cars, easily switching from one fuel source to another. In this case, you can think of glucose as the hybrid car's gas and your fat storage as the battery. When glucose is available, mitochondria use it to make energy while leaving the battery alone to "charge." But when the gas runs out and there's no glucose available, mitochondria can keep running on the stored energy. It's another great system, if—and only if—you operate the car properly.

We are meant to alternate between running on gas and battery power during each twenty-four-hour cycle. During the day, you run on gas. Then at night, you run on battery power. After all, you're not eating in your sleep (at least, I hope not), so there's no new glucose to burn. And you're asleep, so you don't need much energy. Your mitochondria can shift to a slow burn and do repair work using the battery power (in the form of free fatty acids) that was previously stored in your fat cells.

This design not only helps you lose weight but also helps you live when glucose isn't available, either because it's nighttime or because you have to go for a long period without food. The latter was certainly the norm for our ancestors, who had to endure harsh conditions. This hybrid system kept them alive and thriving during lean times.

Mitochondria can lose their metabolic flexibility, however, when we feed our gut buddies the wrong foods, particularly processed foods that offer our gut buddies no nourishment.[1] These foods are the opposite of resistant starch and fiber, which resist digestion and make it to your gut buddies for them to eat instead of you. Processed foods are hyper-digested, leaving nothing for your gut buddies to ferment.

Even worse, many of these foods contain fructose, which is absorbed in the gut and goes straight to the liver, where it is converted into harmful fatty acids such as ceramides and palmitate and released into your bloodstream. Fructose makes up half of the sugar in plain old table sugar (sucrose), which is hiding in nearly every packaged food, as well as in high-fructose corn syrup (hence the name), which has even more fructose than table sugar does. Fructose is also the main sugar in fruit.

So now free fatty acids and glucose are rushing to your mitochondria all at once. This causes a "traffic jam" in your cells, and all that traffic creates ROSs that damage your mitochondria, which eventually lose their metabolic flexibility. This can lead to insulin resistance, obesity, and type 2 diabetes.[2]

The good news is that your metabolic flexibility can be restored when you start feeding your gut buddies not only the right foods but also less often.

CYCLING INTO AND OUT OF HIDING

At all times, your body monitors the amount of energy that is available and uses that information to decide whether to grow—because energy is plentiful—or to pull back and clean house—because times are tough and there isn't much energy available. Your cells communicate this information with one another along a pathway called the mammalian target of rapamycin (mTOR). You can think of mTOR as a sensor for energy availability.

When mTOR senses that there is energy to spare, it activates a growth hormone called insulin-like growth factor 1 (IGF-1), which leads cells to grow. But when mTOR senses that there isn't a lot of energy available, it limits the production of IGF-1, which inhibits growth. By measuring your IGF-1 level, we can therefore assess how much mTOR is or isn't being stimulated in your body.

The problem is that when we overfeed ourselves and our gut buddies, mTOR is constantly stimulated. This can lead to metabolic inflexibility, sure, but it can also cause even bigger problems. For one thing, when IGF-1 tells cells to grow, it doesn't discriminate between healthy cells, old and dysfunctional cells, and cancer cells. They all grow. This is why mTOR, which, again, stimulates IGF-1, contributes to cancer growth and makes cancer less responsive to treatment.[3]

When you give your body a break from producing IGF-1, cancerous cells are repaired or discarded, so they don't have a chance to grow when more energy becomes available and IGF-1 increases again. The same thing happens with other damaged or faulty noncancerous cells that can age you and contribute to a variety of diseases. Bottom line: you don't want mTOR being stimulated all the time. Suppressing mTOR and therefore IGF-1 induces autophagy, improves insulin resistance, increases longevity, and boosts metabolic health.[4]

We are meant to live in a seasonal cycle of growth and regression, stimulating mTOR and putting on weight while we are eating the plentiful food that grows in the spring and summer and then shedding that weight while eating less in the fall and winter. This is when we should be sending the signal to our cells (through low IGF-1) that there isn't much energy available to grow and it's time to focus on cleaning and repairs through autophagy.

Simply put, the daily and yearly cycle should be: IGF-1 goes up, and cells grow and proliferate. Then IGF-1 goes down, and faulty cells are cleaned up or cleared out. When IGF-1 goes up again, there are only healthy cells left to grow. Repeat over and over throughout a long, healthy life.

After all, there are seasons for a reason, and we adapted to eat according to those seasons. Mother Nature never intended for us to have fruit picked before it was ripe, shipped around the world, artificially ripened, and sold in grocery stores—not to mention bred for sweetness. (Cotton candy grapes, anyone?) Now we're consuming these fructose bombs throughout the year. Instead, we are meant to be exposed to ripe, local fruit only during a specific time period each year, when we are in a growth cycle, putting on weight so that we can use the fat stores to make it through the winter season of regression.

Of course, we have become disconnected from this yearly cycle. We are living in an endless summer, constantly sending growth signals to our cells. The result is metabolic inflexibility, weight gain, disease, rapid aging, and the growth of cancer cells, not to mention a dysbiotic microbiome that has adapted to eating that high-fructose-content, unnaturally ripened fruit year-round.

To the contrary, we are meant to experience a cyclical reconfiguration of our microbiome based on what foods are available during each season. The Hadza people of Tanzania are some of the last hunter-gatherers on the planet. During the wet season, they forage for berries and eat more honey, whereas in the dry season, they focus on hunting, so they eat more meat. Throughout the year, they eat fibrous tubers, which are great for their gut buddies. An analysis of the Hadza people's microbiomes across seasons reveals some very interesting things about how the microbiome is meant to adapt to seasonal change.[5]

Fascinatingly, the Hadza people's microbiomes are virtually identical from one wet season to the next and from one dry season to the next. But there are dramatic changes between the dry and wet seasons. During the wet season, for example, the bacteria that ferment fructose grow and multiply. Then they take a back seat in the winter, when fruit is no longer available. This is another one of nature's built-in mechanisms to keep any species from overgrowing, and it protects the Hadza people from metabolic inflexibility, which they do not experience despite eating lots of fruit and honey when it is plentiful.

It's also interesting to note that when you eat this way, many species of bacteria disappear completely during one season and then reappear the next. These are the gut buddies you read about earlier that live in hiding down in the crypts. They are always there, and they can be coaxed out of hiding when you provide them with their favorite foods. But when there's nothing for them to eat, they go back into hibernation.

This offers a ray of light for those suffering from dysbiosis, which unfortunately is most of us. How many of your gut buddies are in hiding right now? By feeding them the right foods at the right times, you can help them reveal themselves and get your ecosystem back into balance.

As I mentioned, in addition to eating seasonally with periods of growth and regression, we are also meant to eat only at certain times throughout the day. The idea that we should be snacking or eating a bunch of small meals all day long is a complete and total myth that has done incredible damage to our health. As you read earlier, your mitochondria need a break at night to clean house and switch over to "battery power." This is also when your melatonin level is high and can act on mitochondria as an antioxidant to clean up ROSs. When your cells are constantly being flooded with fuel, they never have a chance to take a break and complete the cleanup process.

Following a daily cycle of growth and regression also allows the body to produce ketones to act as mitochondrial uncouplers. Normally, we start making ketones eight hours after we last ate. When we give our bodies enough time without food, those ketones help our mitochondria repair themselves and multiply.

Furthermore, we often overlook the fact that digesting and carrying molecules across the gut wall is hard work that requires a huge amount of energy and blood flow. Your gut wall needs downtime to do its repair work and clean up dead or faulty cells, but this does not occur when it is always taxed with the job of digestion. The gut wall is not great at multitasking. Instead of having to digest and clean up at the same time, it's much more efficient to concentrate the digestion workload over a shorter period of time and leave the rest of the day for cleaning. It's another cycle, just like the one we should be following throughout the year.

SUPPRESSING MTOR: THE HARD WAYS AND THE EASY WAYS

Besides following both daily and yearly cycles of growth and regression, there are additional ways to suppress mTOR to protect your health. One is by restricting your intake of animal protein. When mTOR is scanning the body for energy availability, it keeps an eye out for certain amino acids more than others. These are the amino acids that are the most necessary for growth—methionine, cysteine, and isoleucine—which are found primarily in animal protein. These amino acids are deficient in most plant-based proteins. So if you avoid animal protein, you can eat as much as you want and still trick your body into thinking that you're in a regression cycle, so it doesn't trigger production of IGF-1 and stimulate mTOR.

The antiaging drug rapamycin also suppresses mTOR, reduces the rate of aging, and improves age-related diseases.[6] I studied rapamycin, which is actually an antibiotic, as a transplant drug back in the day and found that it helped animals live longer. When my team and many others looked into why that was happening, we found that it was by suppressing mTOR. (Remember, mTOR stands for "mammalian target of *rapamycin*.")

In mice, rapamycin dramatically extends life span[7]—up to threefold![8] It does this in part by altering the structure of the microbiome and reducing the number of bacteria in the small intestine.[9] In studies of mice with multiple sclerosis, rapamycin also alleviated the progression of the disease by inducing autophagy and inhibiting the immune response.[10]

You read earlier that metformin, the most popular treatment option for type 2 diabetes, works by increasing the diversity of your gut buddies. It also increases the abundance of *Akkermansia*, the gut buddy that helps protect the gut wall.[11] Metformin is actually an mTOR suppressor, much like rapamycin. Both drugs extend life span.[12] Ashwagandha, a plant compound that is a natural alternative to rapamycin, has long been used in the Ayurvedic tradition to extend life span. It also suppresses mTOR and creates changes in the gut biome that make you more resilient to stress.[13]

Last but certainly not least, humans have a class of proteins called sestrins that act as antioxidants[14] and regulate the mTOR signaling pathway. Sestrin proteins are expressed in various immune cells.[15] They suppress the inflammatory response, inhibit T-cell immunity, and support white blood cells.[16]

During times of stress, the overexpression of sestrins suppresses mTOR while the sestrin proteins assess cells' viability. If a cell is viable, sestrin proteins can act as antioxidants and clean up any ROSs[17] to keep it healthy. If a cell is only slightly damaged, sestrin proteins activate autophagy while inhibiting apoptosis. This way, the cell is recycled without suffering any damage. If, however, a cell is completely faulty or cancerous, sestrins drive it into apoptosis.[18] They are therefore incredibly important for maintaining and restoring homeostasis.[19]

Sestrins are especially important during the "endless summer" that we are now living in, as they protect us from the effects of overfeeding.[20] When mice that were sestrin deficient were fed a high-fat diet, they developed glucose intolerance, insulin resistance, and fatty liver, which all stemmed from mTOR overactivation.[21] But in another mouse study, sestrins protected mice that were fed a high-fat diet from developing insulin resistance by suppressing mTOR.[22]

By now, I'm sure you're waiting for me to get to the part where this all comes back to your microbiome. Good news, folks, we have now arrived! In fact, this is yet another connection that I was unaware of until I was doing the research for this book, and it boggled my mind when I found it. So here we go.

In 2020, researchers looked at sestrin levels in diabetic mice and found that the mice had significantly lower sestrin levels than did nondiabetic mice. This makes sense, since we know that sestrins protect against metabolic disease. But here's where it gets interesting: when the diabetic mice were given a probiotic (*Lactobacillus delbrueckii*), their sestrin level went up and their serum glucose level went down.[23] Essentially, the probiotic reversed the mice's diabetes by increasing their sestrin expression. But how?

Lo and behold, polyphenols, the favorite food of many of our gut buddies, activate sestrin! Quercetin, a type of flavonoid found in berries, onions, grapes, broccoli, and citrus fruits, strongly activates sestrin and inhibits mTOR in colon cancer cells in a dose-dependent manner.[24] The polyphenol resveratrol also activates sestrin genes and has been shown to protect against metabolic disease.[25] You read earlier that resveratrol also activates SIRT1, which repairs and protects cellular DNA. Well, it turns out that SIRT1 also suppresses mTOR!

Polyphenols regulate mTOR in two ways: through SIRT1 and sestrins. This makes perfect sense, since, as you know, your gut buddies are the ones that activate polyphenols and use them as signaling devices to your cells. In this case, your gut buddies use polyphenols to tell your cells to stop growing, to look out for and kill off faulty cells, to clean up healthy cells, and to repair and protect the cellular DNA. This is yet another way that they use their language to take care of you—if you have the right ones.

MTOR, THE GUT, AND GENE EXPRESSION

There's one more piece to this puzzle, and this one lies in our genes. Though we have more bacterial genes in our bodies than human genes, the human genes we do have are still pretty important. Yet our gut buddies play an enormous role in activating or silencing our human genes.

All of our DNA lies in the nuclei of our cells, wrapped tightly around millions of proteins called histones. Histones basically help organize the overwhelming amount of data in our DNA so it can fit inside the tiny space of a cell's nucleus. Anything that affects these histones therefore impacts your genetic expression. When histone levels change, for instance, certain genes can either be upregulated or repressed.

Interestingly, when fruit flies were treated with rapamycin—which, you will recall, suppresses mTOR—histone levels in intestinal cells went up. This increase in histones led to the reorganization of DNA and activated the genes for autophagy. We now believe that histones are the missing link between rapamycin and longevity. Rapamycin increases histone levels in the gut, which rearrange DNA. This increases autophagy and protects the gut wall, leading to longevity.[26]

But there's even more to the histone story than that. Histone deacetylases (HDACs) are enzymes that remove acetyl groups from the amino acid lysine on histones. That's a lot to make sense of, but all you really need to understand right now is that HDACs allow the histones to spool more tightly, making DNA less accessible. Think of a transcript of instructions that can't be read because it's too tightly coiled. In addition to causing other problems, HDACs allow cancer cells to grow and proliferate. We definitely don't want that. If only there was something that could inhibit the actions of HDACs . . .

There is! HDAC inhibitors have been used in medicine as mood stabilizers, antiepileptics, anti-inflammatories, and antiparasitic drugs. They can also be used to treat cancer and induce tumor cell apoptosis.[27] And guess what is one of the most powerful HDAC inhibitors in the world? Butyrate, the short-chain fatty acid (SCFA) produced by your gut buddies.[28] Though other SCFAs also act as HDAC inhibitors, butyrate is the most effective at inhibiting HDAC activity, stopping cancer cells from proliferating, and stimulating specific gene expression.[29]

This brings me back to the beginning: if you have a robust, stable ecology of gut buddies, you don't need to worry about HDACs, rapamycin, sestrins, or any of the other ways of suppressing mTOR—they will take care of that, and most other things, for you. But first, you have to help them help you by feeding them the right foods at the right times, which you are about to learn how to do.

CHAPTER 10

THE *GUT CHECK* EATING CYCLE

One of the cornerstones of the *Gut Check* program is adjusting your eating schedule to maximize the health of your inner ecosystem. To put it simply, you are going to condense the time during which you eat to a shorter period during the day. Since I started writing about time-restricted eating years ago, it has been branded "intermittent fasting" and become quite popular. Hopefully, this means that the idea is not as intimidating to you as it was to my patients when I started telling them to do it twenty years ago! I promise that you will not starve on this plan. Instead, you will be more nourished than ever.

Trust me when I tell you that this method of eating is what your gut buddies want. And hopefully by now you are more than convinced to give them what they want! Time-restricted eating changes short-chain fatty acid (SCFA) production in the microbiome, leading to an increase in thermogenesis.[1] It also leads your body to produce ketones, which uncouple mitochondria, also increasing thermogenesis and strengthening the gut wall. In case you forgot, thermogenesis promotes weight loss, vitality, and good health.[2]

Plus, to reiterate, the process of digestion is a lot of work for your gut wall. It's when you're digesting that your gut wall suffers the most stress. Moreover, lipopolysaccharides (LPSs) hop onto fat molecule carrier chylomicrons (lipids) to ride into your body during the digestion process.

This is when your gut wall and your immune system are most vulnerable to attack by lectins and other disrupters.

You may be convinced that time-restricted eating is the way to go but still wary of the reality. I understand. This is why my program will help you slowly work your way up (or down, as the case may be) to the ideal eating window. Slowly compressing the period of time during which you eat will help your body become metabolically flexible, and you must be metabolically flexible in order to fully benefit from this (or quite frankly any) eating program.

I'm sorry to say that most of you who are reading this book are likely to be metabolically inflexible. A full half of people who are at a "healthy" weight are metabolically inflexible, as are 88 percent of people whose weight falls into the "overweight" category, and 99.5 percent of those whose weight is classified as "obese." If you fall into one of these demographics, your body isn't able to make the switch from burning glucose to releasing free fatty acids (FFAs) to generate ATP. Even if it could, your high blood insulin level would prevent fat cells from releasing FFAs. If your body can't release FFAs, it can't generate ketones. And without that ability, you're basically lacking a cellular repair system, including in the cells along the all-important gut wall.

Don't let these statistics deter you! By starting slowly and giving yourself time to become metabolically flexible, you will be able to benefit fully from this program. Starting slowly will also reduce your cravings and make the whole thing a lot easier. To that end, you'll begin with an eating window (the amount of time you can eat within any twenty-four-hour period) of twelve hours and work your way down from there. Week by week over the next five weeks, you'll shorten that window to six to eight hours each day, if possible. You'll also stop eating at least three hours before bedtime. Doing so will give your body, your mitochondria, your gut buddies, and your brain the time they need to rest, repair, and regenerate themselves. Perhaps best of all, you need to maintain this compressed eating window only Monday through Friday. Over the weekend, you can be more flexible and eat according to whatever schedule works best for you.

Each week is broken down as follows:

- **Week 1:** You'll start breakfast at 8:00 a.m. and finish your last meal of the day by 7:00 p.m., Monday through Friday. Not bad, right? Once the weekend comes around, you can be more flexible—within reason! No midnight snacking, please. You can eat breakfast when you like, provided you're still following the *Gut Check* program's eating dos and don'ts and list of approved foods.
- **Week 2:** Week 2 is much like week 1 except that this week you will delay your first meal of the day by one hour, breaking your fast (break-fast, get it?) at 9:00 a.m.
- **Weeks 3 through 5:** Each successive week, you'll follow the same basic schedule, but you'll push breakfast back by another hour. This means that during week 3, breakfast will be at 10:00 a.m. Week 4? 11:00 a.m. And so on until you don't have your first bite until noon, condensing your daily eating window to a mere seven hours (noon to 7:00 p.m.).

Note: *I default to 7:00 p.m. as the end of the day's eating window, as that seems to work best for most of my patients and gives them the space to have that three-hour break before bedtime. But the eating windows I've mentioned here are not absolute. You may have different needs due to work or family commitments. Go ahead and break your fast at 9:00 a.m. and stop eating at 4:00 p.m. if that works for you. The idea is to gradually ease your way into a six-to-eight-hour eating window. It doesn't really matter when that window opens and closes.*

Once you've made it through the first five weeks, enjoying the more flexible weekends along the way, you'll quickly see that time-restricted eating can work with just about any lifestyle. And after you complete the initial five-week schedule, you may notice some changes. To start, you will have become more metabolically flexible. Even better, you'll have become accustomed to your condensed eating window, and it won't seem like a hardship anymore.

GUT CHECK TIPS AND TRICKS

The first two weeks of this program can be a little challenging, especially if you're used to eating breakfast first thing in the morning. Trust that when you stick with the program long enough to feel its many benefits, your metabolism will get the right messages from your microbiome and mitochondria, and those hunger pangs will diminish. But if you're finding it hard to manage your hunger, here are a few tips.

First, make sure that you stay well hydrated. When you're getting enough to drink, you feel less hungry. I recommend filtering water to eliminate any waterborne toxins. Consider installing a home reverse osmosis system. There are great countertop units that require no plumbing or hoses made by AquaTru.

If you have a choice of water, my go-to is San Pellegrino sparkling water. Not only is it pH balanced, but it also has the highest purity and sulfur content of any bottled water on the market. Also, the CO_2 that provides the bubbles may benefit blood flow to your organs and your brain.

When choosing a beverage, adding polyphenols is a bonus. You should feel free to indulge your morning caffeine habit with either green or black tea or black coffee. All of these beverages are rich in polyphenols for your gut buddies (and, contrary to popular belief, they don't dehydrate you). Remember, caffeine also works as a mitochondrial uncoupler in its own right. If you don't like caffeine, try decaf or have an herbal tea made with herbs containing polyphenols such as mint. And if you aren't ready to give up your creamer just yet, choose one of the many keto MCT creamers on the market.

To give your water a little flavor and add in the power of fermented polyphenols, try adding a dash of balsamic or apple cider vinegar to make a refreshing sparkler. In addition to all of its other benefits, such as training the immune system,[3] consuming fermented foods (or beverages) increases your levels of conjugated linoleic acid (CLA), a positive health-promoting molecule.[4]

Another trick is to consume MCT oil. Try eating a spoonful of MCT oil (preferably the C8 or C10 varieties, which are more ketogenic) three times a day. You can start with a teaspoon to help you get through those sinking spells and eventually move up to a tablespoon. A note of caution, however: some people experience some gastrointestinal distress after ingesting MCT oil. This is common in many of my female patients. I recommend using it sparingly at first and working your way up to larger amounts. Some people do much better with MCT powders, several of which are available as coffee creamers. The beauty of MCT oil is that it will not break your fast, and of course it generates ketones.

One of the best ways to keep hunger pangs at bay is to increase the amount of prebiotic fiber or fermented foods that you ingest. Again, it's all about keeping your gut buddies happy and satisfied. You can add prebiotic fiber to your meals, but I find that the best way is to just mix a scoop of prebiotic fiber powder such as psyllium or ground flaxseed into some water, add a tablespoon of apple cider or other vinegar, and drink it. If this is not appealing, sweeten it with allulose, a rare true noncaloric sweetener that feeds friendly gut buddies. Because you can't digest any of these things, they won't break your fast, but they will make for the best breakfast your gut buddies could ever ask for. They'll start producing butyrate and send messages telling your brain that they are full and content and no more food is required. Amazing!

Yet another trick is to enjoy an MCT capraccino. No, that's not a typo for "cappuccino"! I use it to refer to the Latin word *capra* for the genus of goat. This tasty beverage, made of coffee and goat milk, will increase uncoupling and is especially helpful if you experience gastrointestinal issues from MCT. This is an extremely tasty and beneficial way to start the day without having to extend your fasting hours.

I also make sure to keep nuts nearby. Whenever a hunger pang feels unbearable, eat one ounce (about a generous handful) of nuts. Salted nuts are best for this. I buy roasted barùkas (baru) nuts and sacha inchi seeds, which have a high polyphenol content, and then grind iodized sea salt onto them.

In general, it's beneficial to increase your salt intake as you work to shorten your eating window. I'll say it again: salt is not your enemy! Ketones actually compete with uric acid, the mischief maker that causes gout, to be excreted from the kidneys. But you can solve this problem by increasing your salt consumption by about a teaspoon a day. Just make sure it's iodized sea salt, as the pink stuff and other varieties of sea salt do not contain iodine.

I'm also fond of electrolyte replacement powders such as those made by the brand LMNT. We even have our own version at Gundry MD. That said, I recommend avoiding most popular electrolyte drinks, even the sugar-free ones, as they tend to be loaded with toxic artificial sweeteners that kill your gut buddies. Not worth it!

When you are hungry, you can also eat a nut or coconut bar. In exciting research published in 2021, my friend and colleague Dr. Valter Longo at the University of Southern California reported that eating a 200-calorie, mostly nut-based bar called a Fast Bar did not interrupt ketone production or raise blood sugar in people who were following a time-restricted eating plan. You heard that right: eating this bar did not interrupt their ketone production from an overnight fast! This is good news. It means that eating a handful of nuts or a nut-based bar will not sabotage your ketone-generating efforts. It will probably keep the hunger demon at bay, too.

Another option to help you deal with your hunger is to take a few capsules or a scoop of preformed ketones in the form of ketone salts or esters. Believe it or not, these supplements are the real deal. They quickly tell your mitochondria that it's time to start uncoupling. Even better, ingested ketones have the same beneficial effects on the gut microbiome as the traditional ketogenic diet. Though ketone salts are relatively easy to find, the esters, quite frankly, are very expensive and taste pretty terrible. It's easier and more palatable for most of my patients to stick with MCT oils or goat and sheep milk products instead.

If all else fails, simply slow the pace. For many of my patients, week 1 is a piece of cake (pardon the pun), but once week 2 rolls around, they feel hungry, cranky, and fatigued. Their inner voice tells them that there's no way they'll make it to week 5.

If this happens to you, stick with the week 1 eating schedule for another week, and then, instead of shortening the eating window by an hour each week, shorten it by half an hour. It may take you more than five weeks to get to the ideal eating window this way, but you'll get there!

You will become comfortable with being uncomfortable, I promise. Then the reverse effect will kick in: you will become comfortable eating this way and uncomfortable not eating this way. It never fails to shock my patients when they come to prefer this way of eating. But it's not surprising, is it? Your gut buddies are in command, and you get to enjoy the ride!

Finally, on some days, you may be struggling and psychologically need to feel fueled and alert. Maybe you have a big presentation or an extra hard workout. In these cases, have your breakfast a bit early, or grab one of those nut bars (or some nuts) on your way out the door. The next day, you can get right back on track. Anytime you make changes to your diet and lifestyle, it is common to experience challenges and setbacks. Go easy on yourself, and do your best to stay the course.

ALLOW SOME FLEXIBILITY

Though I've provided you with a basic schedule, one of the biggest benefits of this form of time-restricted eating is that it offers some flexibility. Provided that you strive for a shortened eating window, you can schedule your meals as best fits your lifestyle. For example, I have patients who are attached to having breakfast at a certain time in the morning. They feel that they need to start the day with something in their belly or they won't be able to function for the rest of the day. If this is you, you have two options: you can grab a nut bar, or you can go for what I like to call the Ramadan option.

If you're familiar with the Islamic faith, you likely know that the ninth month of the Islamic calendar is called Ramadan. During this time, tradition dictates that devout Muslims fast from dawn to sunset and pray. Most families get up early, eat a small breakfast before the sun rises, and then abstain from consuming food and drink until after sunset. The evening meal is when families feast together in celebration.

I mention this because this type of eating schedule can also create a condensed eating window. By fasting for twelve hours during the day and then again for eight hours overnight, those who follow Ramadan traditions are essentially fasting for twenty hours in a twenty-four-hour period.

That said, though this approach works better for some people than the more traditional schedule does, most studies show that it doesn't promote weight loss. I've actually seen this in my patients. They do get most of the health benefits associated with time-restricted eating—there is a drop in their insulin levels and an improvement in their blood work—but about half the people who follow this type of schedule don't lose weight. As I've reviewed these patients' food diaries, I've noticed a trend in the people who maintain their weight: they tend to indulge in a huge meal at night, supplemented with generous portions of dried fruits such as figs and dates. All that fructose likely explains why those extra pounds stay put!

Despite the fact that not everyone will lose weight using the Ramadan approach, there are still benefits to trying it, particularly if weight loss is not on your list of goals. And it's much, much better than giving up entirely. Try eating breakfast, skipping lunch, and then waiting to eat dinner as your last meal of the day. Even better, grab a handful of nuts, a piece of goat or sheep milk cheese, or an approved bar for your breakfast. It's likely that in time, as your metabolism adjusts and becomes more flexible, you'll find that you're more than ready to start pushing out breakfast to a later hour.

Another option that some of my patients appreciate is the one-meal-a-day (OMAD) plan. If you're already engaging in some form of intermittent fasting or you find the first five weeks of the schedule fairly easy, you can turbocharge your progress by eating a single meal each day. This is exactly how I conduct my own time-restricted eating for a large part of each year. From January to June, I eat a single meal between 6:00 p.m. and 8:00 p.m. That's right: I fast for twenty-two hours a day, five days a week, six months of the year.

Though this option may seem extreme (and many of you may doubt that it is even possible!), I can assure you that it is not only possible but, with time, is easier to sustain than you might imagine. And the results are undeniable! When my patients try the OMAD approach, I can immediately see the difference in their blood work. By significantly reducing the time your gut spends digesting food, you can give your mitochondria the rest they need to uncouple and thrive.

I usually drop ten to fifteen pounds every year during this time period—and then, over the next six months, I gain it back. Huh? How stupid is that? Not stupid at all. Think back to our ancestors and their yearly cycle. Our hunter-gatherer forebears had to cope with a limited food supply during the winter and spring. Summer and fall, of course, brought bountiful options. By taking this more seasonal approach, which is deeply engrained in our genes and our gut buddies' genes, your gut buddies and mitochondria will benefit. This is the circadian seasonal rhythm that's been flowing through you and your gut buddies for millions of years. Why mess with success?

If you'd like to give this plan a try, I suggest starting after the five-week marker, when you're eating your first meal at noon and stopping all food intake by 7:00 p.m. During week 6, you can push your first meal to 1:00 p.m.; week 7, to 2:00 p.m. And so on until you reach the two-hour OMAD window by week 11.

If you are considering going OMAD, I must also offer a word of caution. Over and over again, I and others have shown that it is not healthy to consistently stay in a state of ketosis. Humans have never thrived by starving themselves over the long term. Our bodies are designed to benefit from periods of both feast and famine. But famine on its own? That's a nonstarter.

So if you go OMAD, again, lighten up on the weekend and have two meals a day, maybe even three. I mention the Ramadan and OMAD options to, once again, illustrate that there are many ways to successfully embrace time-restricted eating. Provided that you aim for a condensed eating window, you can find a schedule that works for you and your gut buddies.

Okay! Now that you know when you're going to eat, let's take a closer look at *what* you're going to eat.

CHAPTER 11

THE PLANT PARADOX 2.0

It's been more than twenty years since I started treating patients with dietary changes and supplementation rather than resorting to surgery or pharmacologic drugs to restore their health and reverse many common and uncommon ailments. Don't get me wrong; I prescribe drugs when needed, and up until a few years ago, I still performed heart surgery. But thanks to the thousands of patients who volunteered to have blood work done every three months (or whenever a new dietary change or supplement choice was added), a very clear picture began to emerge of which supplements and foods, in which quantities, consumed at which time of the day or year, made a measurable difference in both the patient's blood work and symptoms.

I call what I do *restorative medicine*, paying homage to Hippocrates' belief that the green life-force energy I mentioned earlier would do all the work in restoring patients' health once I identified the dietary culprits and environmental factors keeping that energy suppressed. That initial work resulted in my first book, *Dr. Gundry's Diet Evolution: Turn Off the Genes That Are Killing You and Lose the Weight for Good*, which was published in 2008.

In that book, I wrote about the reversal of several autoimmune diseases, and shortly after its release, more and more people with those diseases started appearing at my clinics. Being a surgeon by training, I told those patients that I didn't know a lot about autoimmune diseases per se. But as a transplant immunology researcher, I sure knew a lot about what the immune system was looking for and how it behaved. I told those patients that if they wanted to "play" and learn with me, I was all in!

With that in mind, if you had asked me twenty years ago what I thought about the term *leaky gut*, I probably would have told you that it was pseudoscience. But thanks to my research background in human evolutionary biology as an undergraduate at Yale University, my knowledge suggested that over possibly millions of years, we had adapted to eating only certain types of foods and that our immune systems might not recognize recent additions to our diet ("recent" meaning within the last ten thousand to five hundred years) as friendly.

As I instructed my patients to remove the "modern" foods that were introduced over the last ten thousand years from their diets, I saw remarkable changes in them and their blood work. Consistently and reproducibly, about 90 percent of the people who followed the program saw their health restored, their autoimmune disease markers and symptoms reversed, their coronary artery disease reversed, their arthritis reversed, their diabetes reversed, their migraines subsided, and so on. It was so remarkable that I published my second book, *The Plant Paradox: The Hidden Dangers in "Healthy" Foods That Cause Disease and Weight Gain*, to critical and popular acclaim in 2017, and the rest, as they say, is history.

Why are that book and the subsequent "Paradox" titles that followed so enduringly popular? Just one reason: the plan works! Not a day goes by that I don't get an email or a message describing a remarkable success story. So you'd think I would have relaxed, patted myself on the back for a job well done, and headed down easy street.

Yet the opposite occurred. I currently see patients six days a week, even on weekends. Why? Let's just say that I'm like a kid in a candy store (bad choice of analogy). I never get tired of seeing my patients "curing" themselves. Now, remember I said that about 90 percent of people who follow my recommendations dramatically improve their health and/or reverse their autoimmune disease. This means that about 10 percent still weren't fully "cured," even though they still often saw improvements. What was I missing?

That's what this chapter is about: what I've learned in the now almost seven years since *The Plant Paradox* was published. As in all of my books, there will be some revelations and recommendations here that you will not like. Sorry. But in many cases, when we discover the missing pieces, a light bulb lights up above my patients' heads, and he or she says things such as "I knew it!" I trust that you will, too.

FOOD SENSITIVITIES AND LEAKY GUT

Thanks to work by Dr. Alessio Fasano, who is now at Harvard, and many others, the science of "leaky gut," or intestinal permeability, is now proven and well established. Not only that, but more and more of the mechanisms, many of which you are now familiar with, are constantly coming to light. And thanks to the National Institute of Health's Human Microbiome Project, which started in 2007 and was completed in 2016, scientists only recently discovered the heretofore unexplored inner "galaxy" of your gut population. Moreover, with each passing year, better tests are being developed to measure and quantify leaky gut and to determine the population of microbes living within an individual's gut. These advances have allowed me (and many others in our field) access to the information I needed to figure out why 90 percent of people restore their health with my program and the remaining 10 percent lag behind.

In my clinics we use Vibrant Wellness (https://www.vibrant-wellness.com/) to test blood and urine for leaky gut, food sensitivities, environmental toxins, mycotoxins, heavy metals, and other things. Early on, a panel of tests was developed to check for celiac disease (the extreme form of gluten intolerance); intestinal permeability using anti-zonulin IgG and IgA, anti-actin IgG and IgA, and anti-LPS IgG and IgA; and antibodies to wheat germ agglutinin (WGA), antibodies to the multiple proteins that constitute gluten, such as gliadins and glutenins, and antibodies to nongluten proteins. (About 25 percent of the proteins in wheat are not gluten.) All these tests together were called the "Wheat Zoomer." (It was named "Zoomer" before the advent of Zoom calls.)

Subsequently, Zoomers were done for antibodies to the multiple proteins in corn, called the Corn Zoomer; multiple antibodies to lectin-containing foods, called the Lectin Zoomer; several antibodies to the various proteins in dairy, including whey, A1 casein, and A2 casein, called the Dairy Zoomer; and antibodies to the proteins in both egg whites and egg yolks, called the Egg Zoomer. Finally, tests were developed to look at IgG and IgA antibodies to two hundred of the most common foods that people eat to determine their food sensitivities.

To be clear, food sensitivities are not the same as food allergies. Allergies are driven by a different immunoglobulin called IgE. We used to do food allergy tests with a hundred or so pinpricks on people's backs, and frankly, I never thought they were much help. But food sensitivities are different. Let me explain.

Ordinarily, the food you eat is digested by your digestive enzymes and your gut microbiome into individual molecules of protein (amino acids), sugar molecules such as glucose or fructose from carbohydrates, and fatty acids. With the exceptions of fats, these molecules are passed through each cell lining your gut and channeled into the large portal vein heading into your liver. If there are gaps in the lining of your gut wall, meaning that you have leaky gut, potentially undigested pieces of food can pass through the wall and be seen as "foreign" by the immune system lining your gut. These guys will say, for example, "What's that piece of broccoli doing in here? I've never seen broccoli before. It's foreign, and I will make an antibody against it, and if I ever see it again, I'll attack it and call in the troops for inflammation." These antibodies are IgG and IgA, not IgE.

What did I learn after doing thousands of tests on my patients over the last ten years, and what does all this mean to you?

Let's start with the Wheat Zoomer, which also is our test for leaky gut. One hundred percent of all patients with autoimmune diseases and markers for autoimmune diseases have markers for leaky gut. Period. All patients with IBS have markers for leaky gut. Period. When tested, all patients with coronary artery disease have leaky gut. Period. Same for diabetes. Need I go on?

What about wheat? Remarkably, when we start working together, 98 percent of tested patients have strong IgG antibodies to wheat germ agglutinin (WGA), gluten, and nongluten components of wheat. Even patients who are "gluten free" and haven't eaten gluten in ten years test positive! How is that possible? Let me explain.

When you were a kid, you got a tetanus shot. Your body made IgG antibodies to the tetanus bacteria so that if you ever stepped on a rusty nail, your immune system would recognize the tetanus bacteria, call in the troops, and kill it. Every ten years or so, you need to get a booster shot because your immunity gradually declines. It needs to be reexposed to the troublemaker so that you will make more antibodies against it.

What the Wheat Zoomer revealed is that nearly all of us have strong antibodies against the various components of wheat. Remember, WGA is such a small lectin that it can cross through the gut wall even without leaky gut, unlike gluten, which actually causes leaky gut. Also remember that WGA occurs only in whole grains. And whole grains have only relatively recently been introduced to our diets. As you know, WGA binds to the glycocalyx on our blood vessels, our blood-brain barrier, our joint surfaces, even the surface of our eyeballs. All of us have antibodies against it, so our body attacks it, along with whatever it's stuck to. Hold that thought; it gets worse.

Why do we have antibodies to wheat, even if we don't currently eat it? That's because with leaky gut, lectins, bacteria, and undigested food are all crossing the gut wall into your body 24/7. The immune system believes that you are under constant attack and keeps all antibodies against foreign substances active, primed, and ready. I see this with every new patient. Literally everyone who shows up to my office, for whatever disease or issue, has antibodies to the various components of wheat, rye, barley, and oats.

That's pretty dire sounding, isn't it? But let's pause a moment for some good news. When my patients seal their leaky gut, all the antibodies to WGA, gluten, and wheat disappear. They're gone. Unmeasurable. The immune system stands down; the gut buddies "teach" the immune system that all is well, they've got your back, and it's time to take their finger off the trigger. Remarkable? Yes. Predictable? Well, Hippocrates knew it twenty-five hundred years ago; I'm just a slow learner.

What about corn? Seventy percent of my patients with leaky gut react to the proteins in corn. Many react to corn as if it were gluten.

Moreover, most corn grown in the United States is GMO and carries a new protein called "cry." The name is apt. Since it is new, most of us develop an antibody to it.

POPEYE WAS WRONG

Now for lectins. As you know, lectins are part of a plant's defense system against being eaten. In *The Plant Paradox* and my subsequent books, the lectin-containing foods list to avoid is extensive, but it is not complete. We now also test for another class of lectins called aquaporins (AQPs). Plants have pores in their leaves that they "breathe" through. Through these pores, water vapor moves in and out, carbon dioxide moves in, and oxygen moves out. These pores are controlled by aquaporins (water pores, get it?).

We also have aquaporins that are nearly identical to plant aquaporins in the wall of our gut, our blood-brain barrier, and the lining around our nerves, called the myelin sheath. Through molecular mimicry, if you develop antibodies to certain plant aquaporins, your immune system can attack your own aquaporins, leading to leaky gut, leaky brain, and leaky nerve sheaths, the last resulting in multiple sclerosis (MS).

Aquaporins are present in most nightshades, including tobacco, bell peppers, tomatoes, potatoes, and soybeans, plus corn, and—drumroll, please—spinach! Sadly, many of my patients with MS, irritable bladder syndrome, and/or brain fog have antibodies to the spinach aquaporin—and they happen to be big spinach eaters. My advice to you readers who have an autoimmune condition is to ditch these aquaporin-containing foods, and that means spinach, as well. Sorry.

Speaking of sorry, a great number of those ten percenters who do everything I've previously recommended and still don't resolve completely test positive for antibodies to all forms of dairy products, including those made with A2 casein, goat milk, and sheep milk, and both egg yolks and egg whites. Again, fear not. When the leaky gut is resolved, most people can reintroduce these foods slowly. In fact, studies suggest that fermentation makes casein much less antigenic,[1] another reason to favor fermented milk products.

WHERE'S THE MEAT?

In my previous books, I noted that all animals have a sugar molecule (sialic acid, for you purists) that coats their glycocalyx. Beef, lamb, pork, and bison have the molecule Neu5Gc, whereas humans, fish, chicken, and other fowl have a similar but not identical one called Neu5Ac. They differ by only one molecule of oxygen, but Neu5Gc is foreign to our immune system. When we eat Neu5Gc-containing foods, we make antibodies to Neu5Gc, and the more Neu5Gc-containing foods we eat, the more antibodies to it we make.[2]

If you've been paying attention, you can probably guess where this is going. Numerous studies have long associated red meat eating with increased risk of developing heart disease, arthritis, and even cancer. We assumed that when our immune system reacts over and over to Neu5Gc (a foreign molecule), it becomes hyperalert and starts reacting to the Neu5Ac lining our blood vessels, our blood-brain barrier, and even our joints, because it's so similar. It appeared that it was another case of molecular mimicry.

Could this be why diets high in red meat are associated with higher rates of coronary artery disease and cancer? Remember, association does not equal causation. Interestingly, cancer cells use Neu5Gc to cloak themselves from the immune system. Neu5Gc is found in tumors. We do not manufacture it in our bodies! So it has to be acquired from our diet. This may explain the association between red meat consumption and cancer.[3]

And that was where the debate stood. Until now. The glycocalyx is where your bloodstream and your white blood cells interact with the surface of your blood vessels, your blood-brain barrier, and your joint surfaces. It's composed of lots of Neu5Ac sialic acid molecules, which are constantly damaged and replenished.

Here's the bad news: Neu5Gc is rapidly absorbed in your small intestine when you eat Neu5Gc-containing foods and can be easily incorporated into your glycocalyx. The more foods with Neu5Gc you eat, the more it replaces the Neu5Ac that is supposed to be there. And the more of these foods you eat, the more antibodies against it you make and—you guessed it—the more you attack the very structure that is supposed to protect your blood vessels, your brain, and your joints.

Want more scary news before I give you an out? The brain abhors Neu5Gc. Even animals that make Neu5Gc won't let it into their brain. They actively keep it out.[4] Yet when Neu5Gc is incorporated into your blood-brain barrier, it is attacked. The now-damaged BBB lets Neu5Gc into the brain, setting the stage for neuroinflammation.[5] Worse, there is strong evidence that Neu5Gc breaking the blood-brain barrier allows the Epstein-Barr virus (EBV) access to the brain during mononucleosis infections. Anti-Neu5Gc antibody levels are very high in patients with multiple sclerosis (MS), suggesting that Neu5Gc is what makes the association of EBV and MS (and, for that matter, chronic fatigue syndrome, or CFS) so intriguing.[6,7]

And guess what? Neu5Gc is a major gut wall disrupter, as well.[8]

So, dear reader, it is no longer a case of association does not equal causation. We now know the cause of the association between red meat and all of the above. Does this mean you can't eat grass-fed, grass-finished beef? Pastured lamb or pork? Bison, deer, elk? That's my general advice. Trust me, that makes this boy from Omaha very sad. But there are several work-arounds.

First, yes, the more Neu5Gc-containing foods you eat, the more antibodies you will make against Neu5Gc and the more it will be incorporated into your vessels, your joints, and your brain. But if you eat a lot less of those foods and a lot more Neu5Ac-containing foods, such as poultry, fish, and shellfish, the more you will replace Neu5Gc with Neu5Ac.[9,10] So can you have your steak and eat it, too? Yes, but be conscious of the fact that every time you eat Neu5Gc-containing foods, you restart the damage that then has to be undone.

Organ meats, particularly liver, contain the highest levels of Neu5Gc.[11] Bone broth? Forget about it! Gut healing? How about gut damaging? But good news, chicken liver and chicken broth both contain Neu5Ac. No wonder those long-lived Italians eat chicken liver pâté and your Jewish grandmother prescribed chicken soup for every ailment!

The second piece of good news is that a recent study showed that your gut buddies and bacteria in general can eat Neu5Gc either in the colon or in foods before it is consumed (during fermentation), so that you will not absorb it and it will not do any damage.[12] How can this be if Neu5Gc is absorbed from the small intestine long before it gets to where the gut buddies can work their magic? The answer has been staring us in the face in the long-lived societies whose people eat these foods. Remember, because meat is such a limited resource, these meat eaters grind up the animal from nose to tail, mix in spices and salt, and allow it to cure, aka ferment. The bacteria eat the sugar molecule Neu5Gc, destroying it. So the answer seems to be eating old-fashioned fermented meats such as cured sausages, chorizo, and salami.

But wait, there's more mischief afoot (or should I say abreast?). The milk from these animals is also loaded with Neu5Gc. Refrigeration is a relatively new invention, yet the practice of milking animals goes back thousands of years. Enter fermentation again. Yogurt, kefir, hard, aged cheeses such as parmesan and pecorino, and cheeses made with raw, unpasteurized milk have significantly reduced Neu5Gc levels. This may explain the paradox in that Swedish study I mentioned earlier showing that drinking milk worsened angina but eating cheese lessened it. The milk had the antigenic Neu5Gc, and the cheese did not. It all starts to add up, doesn't it?

On this plan, I am recommending for the first time that you eat some chicken gristle, some pastured chicken skin, and some sardines and other small fish such as anchovies and herring, bones and all. In addition to their feeding your gut buddies, we don't make enough of many important amino acids such as glutamate and glycine,[13] and the collagen in these contains an amino acid called trans-4-hydroxy-L-proline that we cannot make ourselves and that scavenges for ROSs and is catabolized into glycine in the body![14] It is a good idea for vegetarians and vegans to supplement with vegan collagen. Another form of protein, gelatin, can protect the gut lining, particularly when paired with tannic acid, which is a type of polyphenol.[15]

Small wild fish and shellfish are also good forms of animal protein. The smaller the fish, the better. Sardines, herring, and anchovies, as well as wild salmon and bivalves (clams, oysters, mussels, and the like) are all great sources of omega-3 fatty acids and phospholipids that don't contain much mercury and other heavy metals. Just be sure to avoid "organic" farm-raised salmon, which are fed corn and soybean meal. Thanks to changes in their diet, farmed salmon no longer make DHA and instead produce inflammatory omega-6 fatty acids.

Omega-3 eggs are another good option for many people, but a number of my autoimmune patients do react to the proteins in both the whites and the yolks. When it comes to "meat," you can enjoy smaller amounts (a maximum of four ounces) of the highest-quality poultry and fish you can get. Factory-farmed meats and seafoods, including farmed salmon and shrimp, are fed antibiotics to prevent disease and fatten them up. The antibiotics make their way into the animals' flesh, and when you consume it, they kill off your gut buddies. Focus on organic products, wild seafood, and pasture-raised chicken.

Finally, before you jump on the "Dr. Gundry says I should eat chicken" bandwagon, not so fast. I grew up in Nebraska. Chickens there were farm animals. They went out in the cow pastures every day and scratched through the cow pies, eating the bugs and spreading the manure. Then they'd come back to the henhouse and lay eggs. When an old hen couldn't lay eggs anymore, she became a stewing hen. Her flesh was so gristly and tough that she had to be cooked all day and sometimes all night, breaking down all the collagen that we, and hence our gut buddies, ate. And her meat was loaded with the proper ratio of omega-6 to omega-3 fats, about 3–5 to 1, respectively.

Recent analysis of commercial grocery store chicken shows that the ratio is now 25 to 1. This means that the once anti-inflammatory chicken is now an omega-6 inflammatory bomb. When you do find a good pasture-raised chicken such as the ones you can buy at Lectin-Light Chicken (https://lectinlightchicken.com/), buy the whole bird and use the whole thing, not the boneless, skinless breast that's devoid of all the good stuff such as the spermidine in the skin and the collagen in the tendons.

As for dairy products, you already know that cow milk products in the United States come from a breed of cow that produces milk with a highly inflammatory protein called A1 beta-casein. That's why I recommend you choose goat or sheep milk dairy products or cow milk dairy from southern European (A2 beta-casein) cows. Goat, sheep, and water buffalo cheeses and yogurts are far superior choices, as they are powerful mitochondrial uncouplers, plus the Neu5Gc is eaten in the fermentation process. Lastly, organic heavy cream has no Neu5Gc, so go ahead and get your milk fat globule membranes (MFGMs) from that.

One final note of caution for readers who have been diagnosed with an autoimmune disorder: the 10 percent of patients who don't fully recover on my program usually test positive for reacting to all forms of dairy products, including A2 beta-casein, as well as both egg whites and egg yolks.[16] So if you are reading this book for help, for now, give up all dairy and all eggs. And don't worry, most of you can reintroduce them later as your gut wall heals, your immune system quiets, and your microbiome replenishes and fortifies.

THE TOP OFFENDERS

Let's look at the other most common culprits that show up as sensitive foods in my patients' blood work. Apologies in advance; you may not like some of this. The top twelve offenders, besides the worst lectin offenders already mentioned, are:

- Almonds and almond flour
- White mushrooms
- Ginger
- Pineapple
- Peaches (apologies to Georgia, where I went to medical school)
- White onions
- Lemons
- Bananas
- Nutmeg
- Cinnamon
- Commercial poultry
- Vanilla bean

Now, this list doesn't mean that you personally are sensitive to all or any of these foods, but consider them as additional foods to think about removing from your diet. There are, of course, substitutions. Use brown or other forms of mushrooms; use red or yellow onions instead of white; use nectarines or plums instead of peaches; use limes instead of lemons; use coconut, chestnut, or tiger nut flour instead of almond flour.

To wrap this up, thousands of patient tests later, I can now give you a much clearer and educated program for "good" foods and "bad" foods. There is no need to argue or debate about the dangers or benefits of wheat, rye, barley, and oats, when nearly most people have antibodies against WGA, and gluten. And consuming Neu5Gc in beef, lamb, pork, bison, venison, and other meats will produce antibodies to it and it will be incorporated into the glycocalyx on your blood vessels, blood-brain barrier, and joints, and further worsen leaky gut. This discussion doesn't even touch on the fact that cancerous tumors incorporate dietary Neu5Gc into their cell walls to avoid detection.[17] Oh, my!

Another way to say this is that your grass-fed burger on your organic whole wheat bun with an organic milk shake may be the perfect storm that will ruin your short- and long-term health. On the other hand, eating some traditionally prepared fermented sausages with goat or sheep milk yogurt and some raw cheese for "break-fast" at noon topped off with polyphenols may be the true secret of the "Blue Zones" that's been hiding in plain sight all along.

CHAPTER 12

THE *GUT CHECK* FOOD PLAN

We have finally arrived at the chapter you've been waiting for: the dos and don'ts of the *Gut Check* program along with lists of foods that you can say "yes" to and the foods you should avoid. The biggest change is that overall, you are going to be eating not for yourself but for your gut buddies. When you give them the foods they need to thrive, you will thrive right alongside them. Rest assured that your gut buddies enjoy a wide variety of foods that you will be able to enjoy on the *Gut Check* program, too.

The foods that are best for your gut buddies—and therefore for you—can be broken down into three main categories: prebiotics, probiotics, and postbiotics. I wish I could go back in time and rename these three categories because so many people (understandably) find them confusing, but it's actually quite simple if you break it down this way: prebiotics are food for your gut buddies, such as polyphenols and prebiotic fiber; probiotics are the gut buddies themselves, which are found in yogurt and fermented foods; and postbiotics are the beneficial products that your gut buddies and bacteria and yeasts outside and inside your gut produce, such as the acetate found in vinegar.

Postbiotics are part of the all-important messaging system that your gut buddies use to communicate with your mitochondria and the rest of your body. And of course, you need all three of these to reach a state of homeostasis. You need the gut buddies themselves to do the work for you. They need nourishment to grow, thrive, and do their various jobs. And the postbiotics are the all-important end products that you need to modulate so many of your body's functions.

What exactly do your gut buddies want to eat? Primarily, they need polyphenols and vegetables that contain prebiotic fiber, along with more polyphenols. And they need fermented foods that contain valuable information. If those fermented foods contain polyphenols, even better. These will be the cornerstones of the *Gut Check* program. Don't forget that coffee, dark chocolate, red wine, and champagne all contain polyphenols! Yes, you can indulge in these favorites and more.

You can also enjoy nuts and some seeds, sheep and goat milk dairy products that contain probiotics and postbiotics, pressure-cooked lentils and other legumes, some wild fish, shellfish, and mollusks, and, if you desire, pastured poultry, omega-3 eggs, and, on occasion, fermented meat products such as traditionally cured and fermented sausages. In season, your gut buddies also enjoy low-fructose fruits. As you can see, there is plenty of variety here, as well as flexibility. You can follow this program whether you are vegan, paleo, or anything in between.

Before we get to the specifics, let's take a look at the following.

WHAT WE ARE GOING TO ACCOMPLISH THROUGH THIS PROGRAM

Replenish Your Internal Ecosystem

The foundation of the *Gut Check* program is made of foods that will nourish your gut by feeding your friendliest and most productive gut buddies. This helps them make those important postbiotics and encourages them to send signals to the rest of your body saying that everything is A-okay down in the gut. These foods will also encourage the gut buddies that have gone into hiding to come out and start working for you, too.

It's just as important to eliminate the foods that kill off your gut buddies and/or allow the gang members lurking in your gut to overgrow.

By feeding the good guys instead of the bad guys, you will reach the elusive state of homeostasis. The result will be the type of robust, thriving microbiome that is your birthright.

Fortify the Gut Wall

By eliminating foods that damage the gut wall and adding others—particularly uncouplers that will keep the mitochondria in your gut wall strong and healthy—you will build up your defenses against invaders. The *Gut Check* program will heal your gut wall, dramatically reduce inflammation throughout your body, and help restore your inner ecosystem. Fortifying your gut wall can do amazing things, including reversing severe disease. No matter what state your gut is in now, it can be healed!

Protect the Glycocalyx

To avoid and even reverse symptoms of autoimmunity, you will make sure that your immune system has no reason to attack any of your protective sugar linings, including the blood-brain barrier safeguarding your brain! These linings are the second line of defense that you wouldn't need to worry about if the gut wall hadn't been breached. By healing and protecting them simultaneously, you will see a huge drop in inflammation and autoimmunity.

Uncouple Mitochondria

As you work to restore your microbiome, your gut buddies will be able to send messages through short-chain fatty acids (SCFAs), activated polyphenols, and other compounds telling your mitochondria to uncouple, make more of themselves, and "waste" fuel. This will help fortify the gut wall, protect you from all sorts of diseases, and boost the health of the cells and tissues throughout your body. As another nice effect, by doing a caloric bypass, it will also help you lose weight!

DOS AND DON'TS

The *Gut Check* program builds on my previous Paradox eating regimens, adding the newest information about what foods are best (and worst) for your gut. As in the other programs, you'll aim to avoid gut-busting lectins at all costs, especially that sneaky little lectin wheat germ agglutinin (WGA). You'll add foods to your diet that are especially good for your gut buddies, including plenty of fermented foods and fermented polyphenols that will kick-start their communication process.

To guide your choices and help you incorporate these foods into your meals, I've created a simple list of dos and don'ts.

Do: Eat Polyphenol-Rich Foods

As you know, your gut buddies absolutely adore polyphenols and activate them so that they can act as mitochondrial uncouplers. This helps both your inner ecosystem and the cells throughout your body thrive. We are seeing more and more evidence that eating polyphenols helps protect against intestinal permeability.[1] But of course! When the mitochondria in your gut wall uncouple and make more of themselves, the health and stability of that wall are protected.

Whether you prefer coffee to tea, kiwis to red berries, or endive to kale, the plant kingdom is full of delicious polyphenol-rich options to suit any palate while satisfying and nourishing your gut buddies. However, it's important also to consider the sugar and grain levels of these foods before consuming them. Here's a list of common foods sorted by their polyphenol content, ranked from highest to lowest.[2]

Cloves

Peppermint, dried

Star anise

Cocoa powder

Mexican oregano

Celery seed

Black chokeberries
Dark chocolate
Flaxseed meal
Black elderberries
Chestnuts
Common sage, dried
Rosemary, dried
Spearmint, dried
Common thyme, dried
Lowbush blueberries
Black currants
Capers
Black olives
Highbush blueberries
Hazelnuts
Pecans
Plums
Green olives
Sweet basil, dried
Curry powder
Sweet cherries
Artichokes
Blackberries
Strawberries
Red chicory
Red raspberries
Coffee, filtered
Ginger, dried
Prunes

Almonds
Black grapes
Red onions
Green chicory
Common thyme, fresh
Refined maize flour (masa made from hominy)
Tempeh
Apples
Spinach
Shallots
Lemon verbena, dried
Black tea
Red wine
Green tea
Soy yogurt
Yellow onions
Pomegranate juice (100% juice)
Extra-virgin olive oil
Black beans (pressure-cooked or fermented)
Peaches
Blood orange juice (100% juice)
Cumin
Grapefruit juice (100% juice)
White beans (pressure-cooked or fermented)
Chinese cinnamon
Blond orange juice (100% juice)
Broccoli
Red currants
Pure lemon juice

Apricots

Caraway

Asparagus

Walnuts

Potatoes (pressure-cooked)

Ceylon cinnamon

Parsley, dried

Nectarines

Curly endive

Marjoram, dried

Red lettuce

Quinces

Endive (escarole)

Pumelo juice (100% juice)

Rapeseed (canola) oil, organic

Pears

Soybean sprouts

Green grapes

Carrots

Vinegar

White wine

Rosé wine

Do: Eat Fermented Foods

As you recall, your gut buddies need predigested or prefermented foods that contain intermediary SCFAs before they can do their job of producing those all-important SCFAs for you. This is the only way for you to benefit from the dietary fiber that you're going to eat on the *Gut Check* program.

However, it's important to be careful when choosing which fermented foods to eat. Conventional yogurt is a nonstarter, as it contains tons of sugar along with inflammatory A1 cow milk. Yes, that includes Greek yogurt. Plain sheep and goat milk yogurts are much better choices, as are low-sugar kombucha, raw cheeses, cheese from Italy, France, and Switzerland, apple cider, and any other vinegars, sauerkraut, and kimchi.

Do: Ferment Your Polyphenols

Polyphenols and fermented foods are both essential for your gut, so what happens when you eat fermented polyphenols? Your gut benefits doubly, and so do you. This is because you are consuming those polyphenols already in their active state[3] and your body can put them to use immediately.

Many foods that are high in polyphenols can be fermented, including teas, vegetables, and fruits. Red wine and champagne are sources of fermented polyphenols, as is balsamic vinegar, one of my favorites. It's easier than you may think to incorporate more balsamic vinegar into your diet: you can simply take a shot every morning, add it to your green smoothie, pour it on top of goat or sheep milk yogurt or ice cream, or add it to your salad dressing. (It's much healthier and very easy to make salad dressing at home instead of buying it.) As I shared earlier, one of my favorite tricks is to add a shot of vinegar to sparkling water. I started suggesting adding balsamic vinegar to sparkling water years ago, and this "healthy soda" has now become a trend. I'm glad to see that more people are giving their gut buddies this treat!

One word of caution: many companies are catching on to the power of fermented polyphenols and are selling beverages made of fermented juice. Make sure to take a good look at their sugar content before consuming them!

Do: Get Your Vitamin D

Though this isn't a dietary change per se, vitamin D is so important to the microbiome, particularly the gut wall, that I'm including it here.

Though humans used to get plenty of vitamin D from the sun, we now spend most of our time indoors or covered in sunscreen. As a result, most of us are deficient in vitamin D.

I put almost all of my patients on a pharmacological dose of vitamin D to calm their immune system and recruit the stem cells hiding in the crypts of the microvilli of the gut wall. I have seen time and time again that this repairs the gut wall and reduces inflammation and autoimmunity. Studies are beginning to be published reporting the same thing. In 2022, a study showed that vitamin D helped reduce autoimmunity.[4] Well, of course it did. When your gut wall is intact and your immune system is not on hyperalert, there is no reason for it to attack your body. And in 2023, another study showed that more vitamin D equaled less dementia.[5] Again, this makes perfect sense to me—and hopefully to you now, too.

How much vitamin D should you take? One of the leading vitamin D research groups at the University of California San Diego believes that the average person should take 9,600 IU of vitamin D_3 per day to have safe and therapeutic levels. They have not seen, nor have I, vitamin D toxicity when supplementing with up to 40,000 IU daily.[6] Indeed, many of my autoimmune and leaky gut patients initially require doses of 20,000 IU daily. I aim for levels of 100 to 150 ng/ml on blood tests.

Do: Eat Postbiotic-Producing Foods

When you eat foods that help your buddies create signaling molecules, you're helping them communicate effectively with your immune system and your mitochondria. Cruciferous vegetables such as broccoli, cauliflower, and other sulfur-containing veggies, including onions, garlic, leeks, chives, shallots, and scallions (all part of the allium family), get top billing in the quest to create healthy communication.

Do: Eat Prebiotic Fiber-Rich Plant Foods

When you eat these foods, you can better reach the good bugs that have been cowering in the far reaches of your gut, sidelined because they lack the nutrients they need to do their job. Eating foods that are rich in soluble (and some insoluble) fibers, including tubers, rutabagas, parsnips, radishes, root vegetables, radicchio, endive, okra, artichokes, pressure-cooked beans and legumes, basil seeds, flaxseeds, psyllium, and more, supports the health and reproduction of healthy bacteria in your microbiome. When your gut buddies get the sustenance they need, they'll send messages to your brain saying that their needs are being met. As a result, you will literally feel less hungry and begin to crave more of the foods that are healthy for them—and therefore healthy for you.

One of the best prebiotics is inulin, a type of dietary fiber found in foods such as chicory, asparagus, onions, leeks, and artichokes. As I mentioned earlier, one of the many benefits of consuming inulin is that your gut buddies use it to produce plasmalogens that protect your brain.

Yet another great way to consume prebiotics is to take psyllium husk powder. Recent studies out of Stanford University have shown that psyllium husks are superior to inulin powder in promoting gut buddy diversity. Or try my new favorite, soaked basil seeds. As many of you know, chia seeds are loaded with lectins. But basil seeds provide all the benefits of forming a prebiotic gel without the dangers of chia lectins. Start with a teaspoon a day, mixed into water or not, and work up to a daily tablespoon or even two for maximum effect.

Last, new work out of Italy has shown that the Green Mediterranean diet is superior to a traditional Mediterranean diet in promoting blood vessel flexibility.[7] What is the Green Mediterranean diet? It's a program that's very similar to the one you are about to embark on. In this study, subjects drank four cups of green tea daily and added a generous scoop of duckweed powder (an aquatic "lentil" that has nothing to do with lentils) to their daily smoothie.

Do: Choose Organic Produce

To fully benefit from your plant foods, choose organic products as much as you can. Conventionally grown crops are sprayed with pesticides, including Roundup, which kill off your gut buddies and cause so many of the diseases I've discussed throughout the book. Buying organic produce really does make a difference. Studies confirm that consuming organic plant foods reduces the pesticide load in children.[8]

Do: Eat Whole Foods

Whole foods provide your gut buddies with resistant starches. Because they resist quick digestion, these starches can make it past your small intestine without being absorbed and into your large intestine, where your gut buddies will convert them into important messaging signals such as butyrate. Yams, taro root, sorghum, millet, pressure-cooked rice, and cassava can all be transformed into resistant starches when they are cooked, chilled, and reheated.

Let me be clear: all of these starch-containing plant products are still made of sugars that are bound together in chains by chemical bonds. Cooking breaks the bonds, but you can remake a number of them by chilling the foods after cooking, then reheating them before consuming. This simple, easy trick often makes a big difference, especially in my metabolically inflexible patients.

The more plant material remains in its original form—whole, if you will—the more resistant it will be to digestion (and therefore the more usable by your gut buddies). For example, a yam that was cooked, cooled, and reheated is going to be much more resistant than sweet potato flour or pasta.

I should also add that popular root vegetables, including beets and carrots, contain complex carbohydrates and resistant starches—when they are raw. Unfortunately, if you cook them, even a little, they lose those components.

Do: Eat for Your Mitochondria

Your mitochondria need two special substances to protect them from oxidative stress: the antioxidant melatonin, as well as a range of healthy fats, including phospholipids, a special type of fat molecule found in egg yolks, fish, and other foods. Though your gut buddies can make melatonin from amino acids, a boost of food-sourced melatonin ensures that your mitochondria will be protected. Luckily, getting your melatonin couldn't be easier or more delicious. The foods on the list below contain high levels of melatonin. Please pressure-cook, cool, and reheat the types of rice listed below.

Melatonin-Rich Foods

Melatonin not only cleans reactive oxygen species (ROSs) from your mitochondria; it also works as an uncoupler in its own right. The following foods contain high levels of melatonin (listed from highest to lowest melatonin content).

Pistachios

Mushrooms

Black pepper

Red rice

Black rice

Mustard seeds

Olive oil

Brewed coffee

Red wine

Cranberries

Almonds

Basmati rice

Purslane (the weed the Ikarians eat)

Tart cherries

Strawberries

Flaxseed

Now let's get to healthy fats. Phospholipids and the short- and long-chain omega-3 and long-chain omega-6 fatty acids all help keep mitochondrial membranes in tip-top shape to ensure that ATP production can function without a hitch. In addition, these fats house the uncoupling proteins in the mitochondrial membranes, even as they promote uncoupling themselves. The best sources of the short-chain omega-3 fat alpha-linolenic acid (ALA) are perilla seed oil, flaxseed oil, ahiflower seed oil, and organic canola (rapeseed) oil.[9] Perilla seed oil has been shown to improve microbiome and gut wall function in athletes![10]

Luckily, phospholipids are plentiful in shellfish such as mussels, scallops, clams, oysters, shrimp, crab, squid, and lobster. Omega-3 egg yolks also have generous amounts of short-chain omega-3s, as well as being a rich source of arachidonic acid (AA), a type of long-chain omega-6 fatty acid.[11] Olive oil is also a great source of the fats and polyphenols you need to keep your microbiome in balance.

Do: Drink Alcohol (in Moderation)

I'm ending the list of dos on a high note, but keep reading before you get too excited. Yes, some types of alcohol such as red wine and champagne do contain powerful prefermented polyphenols. But consuming too much alcohol is never a good thing, and you must choose wisely. The grapes used to make some wines are sprayed with gut buddy–destroying pesticides, so always choose organic or biodynamic wine.

It's also important to limit your portions. I recommend that women consume no more than four to six ounces of organic or biodynamic red wine or champagne with dinner. Men can double this amount. If you prefer spirits, you can have one ounce of your favorite dark spirit during your evening meal. Why a dark spirit and not clear alcohol such as vodka or gin? Because dark spirits have been aged in wood casks, and in the process, they've absorbed polyphenols from the wood!

Don't: Eat Lectin-Rich Plant Foods

Though it's important to eat plant foods that are rich in prebiotic fiber, it's equally important to avoid lectin-heavy veggies, grains, and improperly prepared legumes. Remember, lectins are one of the main culprits in causing leaky gut. Once your gut wall has been made permeable, lectins can go on to damage your internal organs and joint tissues, leading to autoimmune disorders including rheumatoid arthritis, Hashimoto's thyroiditis, diabetes, and coronary artery disease.[12]

Perhaps the most damaging type of lectin is WGA, which is small enough to sneak through an intact gut wall and sticks to the glycocalyx linings of your arteries, joints, and/or protecting your brain. There is no place for whole grains containing WGA on the *Gut Check* program.

Remarkably, WGA is in the outer part of wheat, rye, and barley, and a similar compound exists in corn and rice hulls. Think for a moment about all the trouble that people around the world have gone to in order to remove that outer hull. People in Asian cultures eat white rice, not brown. People in cultures who ate corn traditionally soaked it in lime or lye to remove the hulls. That resulted in hominy, which in turn became masa flour and was used to make tortillas.

Where do we even start with the history of trying to make white flour from whole wheat flour? Why is Italian pasta white, not whole wheat? Why? Because the milling process eliminates a major cause of damage to the glycocalyx of blood vessels, joints, and the blood-brain barrier. Remember, WGA is a foreign antigen and inside your body is attacked like a splinter. Is it any wonder, then, that "whole grain goodness" resulted in a wave of arthritis, heart disease, and dementia?

Lectins are also found in nightshade vegetables (including white potatoes, tomatoes, peppers, eggplant, and goji berries), brown rice, beans and lentils, grains, and pseudograins (including amaranth, quinoa, and buckwheat), peanuts, cashews, and chia seeds. The good news is that most lectin-containing foods can be consumed after being pressure-cooked—or fermented! Let those gut buddies eat your lectins for you!

Don't: Eat Processed Foods

Remember, you are no longer eating for you; you are eating for your gut buddies. If you are worried that you'll have trouble giving up your favorite junk foods, keep in mind that you crave those foods because your gut is currently in a state of dysbiosis. The bad guys that thrive on those foods have hijacked the communication systems to your brain, and they are telling it that you want more of them. Once you feed your gut buddies what they need and allows them to grow, healthy communication will be restored, and you will start desiring the foods that are best for them—and for you.

In addition, processed foods are typically full of polyunsaturated omega-6 fats from soybean and corn oils, which destroy your gut buddies' ability to make hydrogen sulfide, a postbiotic gasotransmitter that helps alleviate inflammation in the gut.[13] Processed and fried foods are also hidden sources of trans fats, those now-banned fats that somehow still sneak into our food supply. This type of fat, created during industrial processing, clogs and damages the inner membranes of your mitochondria.

Perhaps even worse, the short-chain omega-6 fat linoleic acid (LA) is the most prevalent fat in most industrial seed oils, including soy, corn, cottonseed, sunflower, safflower, and grape seed oil. When heated, this is transformed into aldehydes, one of the most toxic compounds to your mitochondrial function yet discovered. One way of protecting yourself is to look for the words *contains one or more of these vegetable oils*. If you see those words, put the bottle down and walk away.

Finally, processed foods are chock full of chemicals such as food colorings, artificial sweeteners, and high-fructose corn syrup, all of which damage our microbiomes. A single packet of an artificial sweetener such as sucralose (Splenda) has the power to kill half of the intestinal bacteria living in your gut. Half! Likewise, titanium dioxide, a common additive used as a whitening agent in personal care products such as sunscreens as well as in the powdered sugary topping found on many doughnuts, has been shown to alter the composition of your microbiome and cause inflammation.[14] Your gut buddies need real food. Give them what they want, and no one will get hurt—at least not you!

Don't: Overdo It on the Fruit

Yes, many fruits do contain polyphenols, so at first glance you may assume that eating plenty of fruit is a great idea. Well, you know what they say about what happens when you assume that if some is good, more is better.

The biggest issue with fruit is that it has a high fructose content, which is a mitochondrial toxin and causes changes in the microbiome. It reduces diversity, feeds gang members, and contributes to leaky gut. Fructose also hampers gasotransmitter production[15] and causes defective insulin signaling,[16] leading to insulin resistance; it even causes insulin resistance in kids![17] Are you still sure your kids need a juice box at snack time?

Yes, fruit is "nature's candy," and that's exactly the problem: it basically *is* candy! Historically, humans were exposed to fruit only during a limited time period each year. For that reason, it's fine to eat local fruit in season. I have blackberry and blueberry bushes in my yard, and I enjoy the berries during the six to eight weeks a year they grow. But it's not natural or healthy to have fruit available at any moment year-round.

Even worse, our fruit has now been bred to be even sweeter than is natural, with higher levels of fructose. Modern fruit, with a few exceptions, is not your microbiome's friend. When in season, pomegranate and passion fruit seeds have a very low sugar content and are good choices. Kiwifruit (skin on, please! It's loaded with fiber and polyphenols) and grapefruit are both low in sugar and laden with polyphenols.

Juice, however, should be off limits year-round, as it is basically concentrated fructose. If you are a huge fan of fruit, I recommend something called reverse juicing that I have mentioned in my previous books. This entails juicing your fruit and then, instead of throwing out the pulp and drinking the juice, throwing out the juice, which contains most of the fructose, and eating the pulp instead! It has far less fructose and most of the fruit's polyphenols. You can eat it on its own or stir it into your goat, sheep, or coconut milk yogurt.

Don't: Eat Sugar

Speaking of candy, the majority of processed foods contain highly refined sugars and carbohydrates. This, of course, goes for processed sweets, but it is also true of foods that are marketed to be healthy, such as many bread products. Not only do most of them contain WGA, but the industrial milling process transforms the wheat into rapidly available sugar.

White bread has a glycemic index rating of 100, which is actually higher than that of table sugar!

Even worse, high-fructose corn syrup is present in many prepackaged foods, including energy bars, granola bars, and cookies. When you see the words *corn syrup*, *brown rice syrup*, *all-natural syrup*, *cane syrup*, or *maple syrup* on an ingredients list, you should know that they are all code words for concentrated fructose.

Remember, your gut is home to some potentially harmful bacteria. That's okay. When our gut buddies are well nourished, we don't have to worry about the gang members because the good bacteria will keep them in balance. But unfortunately, the bad guys thrive on saturated fats and simple sugars. When you make these types of foods mainstays of your diet, you give sustenance to the bad bacteria, enabling them to grow in size and number and produce LPSs, causing inflammation, all while starving out all those good bacteria that your health relies on.

Luckily for those of us with a sweet tooth, there are healthy ways to satisfy it. The best alternative for your gut buddies is non-GMO allulose, a naturally occurring sugar that doesn't impact blood sugar or insulin levels. Even better, your gut buddies love the stuff. Allulose causes positive changes to the microbiome and reduces inflammation.[18] When mice are given allulose, they produce more SCFAs.[19] That's a win for both your sweet tooth and your gut buddies.

Don't: Eat Too Much or Too Little Protein

Besides all of the issues with Neu5Gc, the problem with high-protein diets is that they deprive your gut buddies of the fiber and polyphenols they need to thrive and communicate with your mitochondria. In fact, mitochondrial function generally starts to plummet within days of starting a high-protein diet.[20] When you eat too much protein, your gut buddies start producing fewer SCFAs, including butyrate,[21] have no polyphenols to eat and activate, and begin producing more damaging compounds.[22]

Remember, overconsumption of animal protein can also cause your gut buddies to make too much hydrogen sulfide, which damages your cells. Once again, we see the Goldilocks effect: though hydrogen sulfide is generally beneficial, there can be too much of a good thing.

THE *GUT CHECK* FOOD LISTS

Hopefully, this all seems pretty straightforward. To make things even easier for you, I've created the following "yes" and "no" lists. You can also find this information online at DrGundry.com, where you can download these lists in PDF form.

Yes, Please: Postbiotic-Boosting Foods

Cruciferous Vegetables

Arugula
Bok choy
Broccoli
Brussels sprouts
Cabbage, green and red
Cauliflower
Collards
Kale
Kimchi
Kohlrabi
Napa cabbage
Sauerkraut (raw)
Swiss chard
Watercress

Other Postbiotic-Boosting Vegetables

Artichokes

Asparagus

Bamboo shoots

Basil

Beets (raw)

Carrot greens

Carrots (raw)

Celery

Chicory

Chives

Cilantro

Daikon radishes

Endive

Escarole

Fiddlehead ferns

Frisée

Garlic

Garlic scapes

Ginger

Hearts of palm

Horseradish

Jerusalem artichokes (sunchokes)

Leeks

Lemongrass

Mesclun

Mint

Mizuna

Mushrooms

Mustard greens

Nopales (cactus paddles; if you can't find them locally, buy them online)

Okra

Onions

Parsley

Parsnips

Perilla

Puntarelle (an Italian chicory)

Purslane

Radicchio

Radishes

Red- and green-leaf lettuces

Romaine lettuce

Rutabaga

Scallions

Sea vegetables

Seaweed and algae

Shallots

Spinach (warning: contains an aquaporin lectin)

Water chestnuts

Fruits That Act like Fats

Avocado (up to a whole one per day)

Olives, all types

Oils

Avocado oil

Black seed oil

Canola oil (non-GMO, organic only!)

Coconut oil (has some effect)

Cod liver oil (those flavored with lemon and orange have no fish taste)

Flaxseed oil (high lignan)

Macadamia oil (omega-7)

MCT oil

Olive oil, extra-virgin first cold pressed

Perilla oil (contains lots of ALA and rosemarinic acid, both uncouplers)

Red palm oil

Rice bran oil

Sesame oil, regular and toasted

Walnut oil

Nuts and Seeds

Up to ½ cup per day.

Almonds (only blanched or marcona)

Barùkas (baru) nuts

Basil seeds

Brazil nuts (in limited quantities)

Chestnuts

Coconut meat (but not coconut water)

Coconut milk/cream (unsweetened full-fat canned)

Coconut milk (unsweetened dairy substitute)

Duckweed powder

Flaxseeds (ground fresh)

Hazelnuts

Hemp protein powder

Hemp seeds

Macadamia nuts

Milkadamia creamer (unsweetened and not the milk)

Nut butters (if almond butter, preferably made with blanched almonds, as almond skins contain lectins)

Pecans

Pili nuts

Pine nuts

Pistachios

Psyllium seeds/psyllium husk powder

Sacha inchi seeds

Sesame seeds

Tahini

Walnuts

"Energy" Bars

Limit to one per day, please.

Adapt bars: coconut, chocolate

Fast Bar

Gundry MD bars

Keto Bars: almond butter brownie, salted caramel, lemon poppy seed, chocolate chip cookie dough, mint chocolate, dark chocolate coconut almond, chocolate-covered strawberry

Keto Krisp: chocolate mint, almond butter, chocolate raspberry, almond butter chocolate chip, almond butter and blackberry jelly

Kiss My Keto: cookie dough, chocolate coconut, birthday cake

MariGold: ChocoNut, Pure Joy, espresso, ginger coconut

Primal Kitchen: almond spice, coconut lime

Rowdy Bars: keto chocolaty cookie dough

Stoka: vanilla almond, coco almond

Processed Resistant Starches

Can be eaten every day in limited quantities; those with prediabetes or diabetes should consume only once a week on average.

Barely Bread bread and bagels (only those without raisins)

Bread SRSLY sourdough nonlectin bread and rice-free sourdough rolls

Cappello's fettuccine and other pasta

Crepini egg thins

Fullove Foods keto hemp and linseed bread

Julian Bakery Paleo wraps (made with coconut flour), paleo thin bread, almond bread, sandwich bread, coconut bread

Lovebird Cereals (unsweetened only)

ONANA tortillas

Positively Plantain tortillas

The Real Coconut coconut and cassava flour tortillas and chips

Siete chips (be careful here; a couple of my "canaries" have reacted to the small amount of chia seeds in the chips) and tortillas (only those made with cassava and coconut flour or almond flour)

Superbloombakery.com (breads)

Terra cassava, taro, and plantain chips

Thrive Market organic coconut flakes

Tia Lupita grain-free cactus tortillas

Trader Joe's jicama wraps, plantain chips

Resistant Starches

Eat in moderation. People with diabetes and prediabetes should initially limit these foods.

Baobab fruits

Cassava (tapioca)

Celery root (celeriac)

Glucomannan (konjac root)

Green bananas

Green mangoes

Green papayas

Green plantains

GundryMD Popped Superfood Crisps

Jicama

Millet

Parsnips

Persimmon

Rutabagas

Sorghum

Sweet potatoes or yams

Taro root

Tiger nuts

Turnips

Yucca

"Foodles" (Acceptable "Noodles")

Diabetic, prediabetic, and insulin-resistant people should use with extreme moderation except for konjac-based noodles and rice or hearts of palm noodles or rice.

Big Green millet and sorghum pastas

Edison Grainery sorghum pasta

GundryMD konjac shirataki noodles

GundryMD sorghum spaghetti

Jovial cassava pastas

Kelp noodles

Konjac noodles

Miracle Noodle kanten pasta

Miracle Rice

Natural Heaven hearts of palm spaghetti and lasagna noodles

Palmini hearts of palm noodles

Shirataki noodles

Slimdown360 sweet potato pasta elbow macaroni

Trader Joe's cauliflower gnocchi

Wild-Caught Seafood

Use with caution owing to its microplastics content. Consume up to four ounces per day.

Alaskan salmon (contains very few microplastics)

Anchovies

Calamari/squid

Canned tuna

Clams

Cod

Crab

Freshwater bass

Halibut

Hawaiian fish, including mahimahi, ono, and opah

Lake Superior whitefish

Lobster

Mussels

Oysters

Sardines

Scallops

Shrimp (wild only)

Steelhead

Trout

Pastured Poultry

Consume four ounces per day.

Chicken

Duck

Game birds (pheasant, grouse, dove, quail)

Goose

Heritage or pastured turkey

Ostrich

Pastured or omega-3 eggs (up to 4 daily)

Pastured chicken or turkey jerky (low-sugar versions)

Meat

100 percent grass-fed and grass-finished; consume four ounces per week; see the previous chapter.

Beef

Bison

Boar

Elk

Pork (humanely raised, including prosciutto, Ibérico ham, Cinco Jotas ham)

Traditionally fermented sausages (good news: they contain no Neu5Gc)

Plant-Based Proteins and "Meats"

Duckweed powder

Flaxseed protein powder

GundryMD ProPlant protein shakes

Hemp protein powder

Hemp tofu

Hilary's root veggie burger (hilaryseatwell.com)

Just eggs (ju.st)

Perfect Day vegan whey and casein

Pressure-cooked lentils and other legumes (canned, such as Eden or Jovial brand) or dried, soaked, then pressure-cooked (use an Instant Pot)

Protein isolates of and/or hydrolyzed pea, soy, or other similar bean powders (not the same as regular pea protein, soy protein, lentil protein, chickpea protein—buyer beware!)

Quorn products: only meatless pieces, meatless grounds, meatless steak-style strips, meatless fillets, meatless roast (avoid all others as they contain lectins/gluten)

Textured vegetable protein (TVP)

Polyphenol-Rich Fruits

Limit to one small serving on weekends and only when that fruit is in season, or unlimited with "reverse juicing." The best options are pomegranate and passion fruit seeds, followed by raspberries, blackberries, strawberries, then blueberries, grapefruit, pixie tangerines, and kiwifruits. (Eat the skin for more polyphenols.)

Apples

Apricots

Blackberries

Blueberries

Cherries

Citrus, all types (no juices)

Cranberries (fresh)

Guava

Kiwis

Nectarines

Papaya

Passion fruit

Peaches

Pears, crispy (Anjou, Bosc, Comice)

Persimmon

Plums

Pomegranates

Raspberries

Starfruit

Strawberries

Dairy Products and Replacements

Aged cheeses from Switzerland

Aged "raw" French/Italian cheeses

Buffalo milk butter (available at Trader Joe's)

Buffalo milk mozzarella: mozzarella di bufala (Italy), Buf Creamery (Uruguay)

Coconut yogurt (plain)

French/Italian butter (limit)

Ghee (grass-fed) (limit)

Goat milk and sheep milk kefir (plain)

Goat milk cream flakes: Mt. Capra

Goat milk cheeses: feta, Brie, mozzarella, cheddar

Goat milk ghee (limit)

Goat yogurt (plain)

Kite Hill ricotta cheese

Lavva plant-based yogurt

Organic heavy cream

Organic sour cream

Parmigiano-Reggiano cheese

Sheep milk cheeses: pecorino romano, pecorino sardo, feta, Manchego

Sheep milk yogurt (plain)

So Delicious vegan mozzarella, cream cheese

Herbs, Seasonings, and Condiments

Avocado mayonnaise

Coconut aminos

Fish sauce

Herbs and spices (all except red pepper flakes)

MCT mayonnaise

Miso paste

Mustard

Nutritional yeast

Pure vanilla extract

R's KOSO, other KOSOs

Sea salt (iodized)

Tahini

Vinegars (apple cider vinegars, Bliss vinegars, Sideyard Shrubs vinegars, others)

Wasabi

Flours

Almond (blanched, not almond meal)

Arrowroot

Cassava

Chestnut

Coconut

Coffee fruit

Grape seed

Green banana

Hazelnut

Millet

Sesame (and seeds)

Sorghum flour

Sweet potato

Tiger nut

Sweeteners

Allulose (By far the best option! Look for non-GMO)

Erythritol (Not as evil as some think. Swerve is my favorite, as it also contains oligosaccharides)

Inulin (Just Like Sugar is a great brand)

Local honey and/or manuka honey (very limited!)

Monk fruit (luo han guo; the Nutresse brand is good)

Stevia (SweetLeaf is my favorite; also contains inulin)

Xylitol

Yacon syrup (Super Yacon Syrup is available at Walmart; Sun-Food Sweet Yacon Syrup is available on Amazon)

Chocolate and Frozen Desserts

Coconut milk dairy-free frozen desserts (the So Delicious blue label, which contains only 1 gram of sugar; but be careful: may contain pea protein)

Dark chocolate, unsweetened, 72% cacao or greater (1 ounce per day)

Enlightened ice cream

Keto ice cream: chocolate, mint chip, sea salt caramel

Killer Creamery ice cream: Chilla in Vanilla, Caramels Back, and No Judge Mint

Mammoth Creameries: vanilla bean

Natural (nondutched) cocoa powder, unsweetened

Nick's vegan ice cream

Rebel Creamery ice cream: butter pecan, raspberry, salted caramel, strawberry, vanilla

Simple Truth ice cream: butter pecan and chocolate chip

Beverages

Champagne (6 ounces per day)

Coffee

Dark spirits (1 ounce per day)

Hydrogen water

KeVita low-sugar kombucha (coconut, coconut mojito, for example), other low-sugar kombuchas

Red wine (6 ounces per day)

Reverse osmosis filtered water (AquaTru)

San Pellegrino or Acqua Panna water

Tea (all types)

No, Thank You: Major Lectin-Containing Foods

Refined, Starchy Foods

Bread

Cereal

Cookies

Crackers

Pasta

Pastries

Potato chips

Potatoes

Rice

Tortillas

Wheat flour

Whole wheat flour

Grains, Sprouted Grains, Pseudograins, and Grasses

Barley (cannot be pressure-cooked)

Barley grass

Brown rice

Buckwheat

Bulgur

Corn

Corn products

Corn syrup

Einkorn

Kamut

Kasha

Oats (cannot be pressure-cooked)

Popcorn

Quinoa

Rye (cannot be pressure-cooked)

Spelt

Wheat (pressure cooking does not remove lectins from any form of wheat)

Wheatgrass

White rice (except pressure-cooked white basmati rice from India, which contains high-resistant starch; American white basmati does not)

Wild rice

Sugar and Sweeteners

Agave

Coconut sugar

Diet drinks

Granulated sugar (even organic cane sugar)

Maltodextrin

NutraSweet (aspartame)

Splenda

Sweet 'N Low (saccharin)

Sweet One and Sunett (acesulfame-K)

Vegetables

Most of these can be made safe foods with pressure cooking; marked with an (*).

All beans* (including sprouts)

Chickpeas* (including as hummus)

Edamame*

Green/string beans*

Legumes*

Lentils*

Pea protein (unless pea protein isolate or hydrolysate)

Peas*

Soy*

Soy protein (unless soy protein isolate or hydrolysate)

Sugar snap peas

Tofu*

Nuts and Seeds

Almonds, unblanched

Cashews

Chia seeds

Peanuts

Pumpkin seeds

Sunflower seeds

Fruits

Some of these we call vegetables.

Bell peppers

Chili peppers

Cucumbers

Eggplant

Goji berries

Melons (any kind)

Pumpkins

Squash (any kind)

Tomatillos

Tomatoes

Zucchini

Milk Products That Contain A1 Beta-Casein

Butter (even grass-fed), unless from A2 cows, sheep, goats, or buffalo

Cottage cheese

Cow milk

Cow milk cheese from American cows

Frozen yogurt

Ice cream (most)

Kefir from American cows

Ricotta

Yogurt (including Greek yogurt)

Oils

All "partially hydrogenated" oils

Corn

Cottonseed

Grape seed

Peanut

Safflower

Soy

Sunflower

"Vegetable"

Seasonings

Ketchup

Mayonnaise (unless MCT or avocado)

Red pepper flakes

Soy sauce

Steak sauce

Worcestershire sauce

Now you know about all the foods that your gut buddies do and don't want to eat. Perhaps it's a good time to remind yourself why you are doing all of this. You've learned that the complex galaxy living in your body controls just about everything that's going to happen to you. This galaxy far and away outnumbers your human cells and human genetic material. Yet you've also read about all of the ways we've ignored, abused, and outright killed these microbes. They're mad as hell, and they're not going to take it anymore.

The good news is that you have the power to rectify the situation. You can undo the damage that's been done to your inner ecosystem, and it's not all that difficult. If you still don't think that's necessary, all I have left to tell you is a quote from yet another one of my favorite movies, *Dirty Harry*: "You've got to ask yourself one question: Do I feel lucky?"

Well, do you, punk? What do you say? I say, let's eat!

RECIPES

Condiments

Miso Caesar Dressing

This is not a classic Caesar dressing (which is Gundry approved if made using olive oil), but in my opinion, it's a fun, fermented upgrade, thanks to the deep, savory flavor of the umami-rich miso paste. Serve over Not-Quite-Classic Caesar Salad (page 212) or as a dipping sauce for your favorite fritters or even grass-fed grilled beef skewers.

Serves 2.

1 tablespoon white miso paste
1 pasture-raised egg yolk*
1 clove garlic
1 anchovy (optional and delicious)
1½ tablespoons Worcestershire sauce
1 tablespoon Dijon mustard
¼ cup grated parmesan cheese
Juice of ½ lemon
3 tablespoons extra-virgin olive oil*

Combine the miso paste, egg yolk, garlic, and anchovy, if using, in a food processor, and pulse until well combined.

*If you're not comfortable using raw egg yolk, omit the egg yolk and olive oil and instead fold in ¼ cup Gundry-approved mayonnaise.

Add the Worcestershire sauce, mustard, parmesan cheese, and lemon juice, and process until smooth.

With the motor running, stream in the olive oil until a creamy dressing is formed.

> To make this recipe vegan, omit the anchovy, use ¼ cup unsweetened coconut yogurt instead of the egg yolk and olive oil, and swap out the parmesan cheese for nutritional yeast.

Black Garlic Aioli

A traditional aioli couldn't be simpler—it's mayo with garlic. This twist on the classic adds an addictive natural sweetness thanks to the fermented black garlic. It takes a little bit of time to come together, even in the food processor, but it's well worth the wait! Try this on Miso Lamb Burgers with Mint Sauce (page 234), or Sauerkraut Fritters with Leafy Greens (page 225).

Makes 1 cup.

2 cloves white garlic
8 cloves black garlic
1 pasture-raised egg yolk
2 tablespoons white vinegar*
½ teaspoon salt
¾ cup extra-virgin olive oil

In a food processor, pulse together the white and black garlic until a smooth paste is formed, scraping down the sides intermittently.

Add the egg yolk, vinegar, and salt to the mixture, and pulse until incorporated into the paste.

With the food processor running, slowly pour the olive oil into the garlic mixture, stopping occasionally to scrape down the sides. Go *slowly*; it should take a couple minutes to get all the olive oil worked into the garlic.

Continue processing until the mixture is thick and creamy.

*This recipe also works beautifully with balsamic vinegar to highlight the sweetness of the black garlic.

Not-Quite-Classic Caesar Salad

Truth be told, I love a classic Caesar salad—but sometimes I want something a little different. That's why I created not one but three fun variations for you—from a mostly classic, miso-infused version to a cabbage version, which is a cross between a salad and a slaw, to a grilled version, which is perfect with a nice barbecue.

Serves 2.

Classic-ish Caesar Salad

1 small head romaine lettuce
Miso Caesar Dressing (page 209)
¼ cup shredded parmesan cheese
¼ cup toasted walnuts

Chop or tear the lettuce into bite-sized pieces and place in a large salad bowl. Toss with the dressing and parmesan cheese. Serve topped with walnuts.

Cabbage Variation

Miso Caesar Dressing (page 209)
Zest and juice of 1 lemon
½ head red cabbage, shredded
1 bulb fennel, shredded
¼ cup shredded parmesan cheese
¼ cup toasted hazelnuts

Combine the dressing with the zest and juice of the lemon. Combine the cabbage and fennel in a large salad bowl. Toss with the dressing and parmesan cheese. Top with the hazelnuts and serve.

Grilled Kale Variation

2 bunches lacinato kale, whole leaves
2 tablespoons extra-virgin olive oil
Miso Caesar Dressing (page 209)
¼ cup shredded parmesan cheese
¼ cup toasted pine nuts

Heat a grill or grill pan to medium-high heat. Toss the kale with the olive oil. Grill for 1 to 2 minutes per side, until the leaves are just charred. Let cool, and cut away the tough stems. Cut into bite-sized pieces. Toss with the dressing and parmesan cheese. Top with the pine nuts and serve.

Kimchi Carbonara with Sorghum or Sweet Potato Pasta

I find that sorghum spaghetti is a beautiful sub for traditional (and lectin-loaded) wheat pasta. If you're vegan, look for a vegan kimchi and use half a can of coconut cream instead of the eggs and cheese for a slightly different—but equally delicious—creamy pasta.

Serves 4.

1 pound sorghum spaghetti
4 tablespoons sesame seed oil
1 small shallot, minced*
6 medium cloves garlic, minced
1 tablespoon minced fresh ginger
¼ cup dry white wine**
½ cup diced cabbage kimchi
¼ cup kimchi juice
3 large, room temperature pasture-raised egg yolks
½ cup grated parmesan cheese, plus more for serving
1 cup thinly sliced spinach
Salt and pepper to taste

Cook the spaghetti in a large pot of water, according to the cooking instructions on the package. Drain, reserving 1 cup of the pasta water.

In a large skillet, heat the oil over medium heat. Add the shallot, and cook for 2 to 3 minutes, until translucent. Add the garlic and ginger and cook for 1 minute, until very fragrant. Deglaze the skillet with the wine or broth, stirring to loosen the browned-on bits. Add the kimchi to the skillet and cook, stirring regularly, until slightly wilted, 1 to 2 minutes. Add the cooked pasta and toss.

*If shallots aren't readily available in your grocery store, use ¼ red onion instead.
**If you prefer not to cook with wine or don't have an open bottle, feel free to use lectin-free broth, and add a squeeze of lemon.

Remove from the heat, then add the kimchi juice, egg yolks, and parmesan cheese and toss vigorously to coat the pasta. If the pasta seems dry, add some reserved pasta cooking water. Divide among four bowls and top with the spinach before serving. Add salt and pepper to taste.

Kimchi Pancakes

Kimchi pancakes (kimchijeon) are one of my favorite Korean foods—but they're often made with wheat flour. By swapping out the wheat flour for lectin-free tapioca starch (do *not* swap in cassava flour), you get an extra crispy, delicious savory treat that's good hot, at room temperature, or cold.

Serves 4.

FOR THE PANCAKES:

½ cup cabbage kimchi
3 scallions, finely chopped
½ teaspoon iodized sea salt
½ teaspoon garlic powder
1 shallot, minced
1 tablespoon kimchi juice
4 tablespoons tapioca starch
3 large pasture-raised eggs
1 teaspoon gochujang (fermented Korean chili sauce)
2 tablespoons perilla or sesame oil, divided

FOR THE SAUCE:

2 tablespoons coconut aminos
2 tablespoons rice wine vinegar
1 tablespoon toasted sesame oil
1 small clove garlic, crushed

In a large bowl, combine the kimchi, scallions, salt, garlic powder, and shallot. Set aside.

In a separate bowl (or a blender to save time), combine the kimchi juice, tapioca starch, eggs, and gochujang. Blend until smooth. Fold the batter into the kimchi mixture.

Heat a large skillet over medium-high heat, and add 1½ tablespoons of the oil.

When the skillet is hot, add ½ cup of the pancake mixture and spread onto the pan as thinly as possible. Reduce the heat to medium low and cook for 2 minutes, then flip and cook for an additional minute.

Set on a wire rack to cool, and repeat steps with the remaining batter and oil.

Make the sauce by combining all the ingredients in a small bowl and mixing.

Slice the pancakes into wedges, then serve them with the sauce.

Cabbage Slaw with Miso-Sesame Vinaigrette

I can't resist a savory slaw—but in my opinion, the best slaw needs a little bit of sweetness, too. To that end, I've worked crispy green pear and a little allulose into this tangy, sesame-forward slaw. I don't like my slaw heavily dressed, but if you prefer yours with a thicker dressing, simply double the dressing recipe and add more to your liking.

Serves 4.

2 cups thinly sliced red cabbage
1 red onion, thinly sliced
2 cups thinly sliced kale, ribs removed
½ teaspoon salt
1 crisp green pear, shredded (optional)
2 tablespoons toasted sesame seeds
1¼ cups unsweetened coconut milk yogurt
¼ cup tahini
2 tablespoons toasted sesame oil
2 tablespoons miso paste
1 teaspoon allulose
Juice of 1 lemon

In a large bowl, massage the cabbage, onion, and kale with salt until tender. Toss in the pear, if using, and the sesame seeds. Set aside.

In a blender, combine the yogurt, tahini, sesame oil, miso paste, allulose, and lemon juice. Blend until smooth and creamy, thinning with water as needed.

Toss the vegetables with the dressing until evenly coated.

Serve chilled or at room temperature.

Kraut and Avocado Breakfast Bowl

Some weekends call for a hearty brunch—and this delicious breakfast bowl is just the ticket. The sauerkraut adds a beautiful tang to balance out the creamy avocado and eggs. I often skip the millet for a less starchy breakfast, but it's great either way.

Serves 2.

1 cup cooked millet
1½ tablespoons extra-virgin olive oil, divided
2 cloves garlic, minced
1 bunch kale, shredded
½ teaspoon salt
1 cup drained sauerkraut
2 pasture-raised eggs
1 avocado, minced
Juice of 1 lemon
Fermented hot sauce, to taste

Divide the millet into two bowls.

In a large skillet, heat half the oil over medium heat. Add the garlic and cook for 1 to 2 minutes, until fragrant. Add the kale and salt and cook for 3 to 4 minutes, until the kale is wilted. Add the sauerkraut and cook for 2 minutes, until heated through. Divide into the bowls, then wipe the pan clean.

Heat the remaining oil over medium-high heat and cook the eggs to your liking.

Put 1 egg on each bowl, along with half an avocado. Squeeze the lemon juice over the top, and add hot sauce to taste.

Pressure-Cooked Chickpeas and Tomatoes with Sauerkraut and Crispy Garlic

Most people don't know this, but I *love* tomatoes and beans—they're honestly some of the most delicious foods on the planet. But you *have* to pressure-cook them to minimize the impact of their lectins. I suggest serving this dish over lectin-free pasta or with lectin-free bread.

Serves 4.

¼ cup sesame seed oil
1 medium red onion, diced
4 cloves garlic, minced
2 tablespoons fresh rosemary, minced
1 tablespoon fresh oregano, minced
½ teaspoon iodized sea salt
2 tablespoons tomato paste
1 28-ounce can peeled, seeded tomatoes*
3 cups lectin-free chicken or vegetable broth
2 cups dry cannellini beans, soaked in three changes of water over the course of twelve hours** (In a hurry? Use Eden or Jovial brand pressure-cooked beans)
2 cups drained sauerkraut
½ cup crumbled goat cheese

Heat your Instant Pot using the sauté button. Add the oil and onion and cook for 3 to 4 minutes, until the onions start to appear translucent. Add the garlic, rosemary, oregano, and sea salt and sauté, stirring occasionally, for 1 to 2 minutes, until the garlic is very fragrant. Add the tomato paste and cook for 1 minute.

* I find it easiest to use peeled plum tomatoes and simply chop them in half, then scoop out and discard the seeds.

**If you don't have time to soak the beans overnight, double the cooking time—and rinse them in multiple changes of water and soak for at least an hour before cooking.

Turn off the heat and add the tomatoes and broth. Scrape the bottom of the Instant Pot to loosen any cooked-on bits of food.

Strain the beans and add them to the Instant Pot. Secure the lid of your Instant Pot, making sure that the top dial is flipped to seal. Cook on high pressure for 20 minutes. Let the pressure release naturally. Once the pressure has released, remove the lid and stir in the sauerkraut.

Top with crumbled goat cheese and serve.

Miso-Glazed Turnips

This miso-glazed turnip recipe was inspired by a recipe I tried in a Japanese restaurant in Los Angeles. It has a delicious sweet-and-savory flavor that's perfect as a side dish for a special meal, and here's the best part: it's shockingly easy to make! No mint in the house? Use parsley or basil. It's tasty no matter what!

Serves 4.

3 tablespoons white miso paste
2 tablespoons grass-fed butter or avocado oil
1½ teaspoons allulose
1 tablespoon coconut aminos
1 pound white turnips, greens removed, scrubbed well, and cut into wedges
Juice of ½ lemon
¼ cup fresh mint, minced

In a large sauté pan, heat the miso, butter or oil, allulose, and coconut aminos over medium heat. Add the turnips and toss to combine. Add just enough water to cover the turnips and turn the heat to medium high.

Bring to a boil, turning the turnips occasionally, until most of the liquid is evaporated and the turnips are fork tender, about 20 to 25 minutes. If the liquid evaporates before the turnips are tender, add more water.

Once the liquid has cooked down, cook for 5 to 7 minutes, stirring occasionally, until the turnips are golden brown.

Serve with a squeeze of lemon juice and a dusting of fresh mint.

Beet Salad with Miso and Black Garlic Dressing

A lot of people find the fact that beets are included on the Gundry plan confusing, and I get it: sometimes they're okay, sometimes they're not. In a raw preparation such as this, where the sugar hasn't had a chance to cook down and concentrate, they're perfectly fine—*especially* when combined with nutrient-rich seasonal fruits, fresh vegetables, and a tangy fermented dressing.

Serves 4.

FOR THE DRESSING:

2 cloves black garlic
1 clove white garlic
1 tablespoon white miso paste
3 tablespoons balsamic vinegar
⅓ cup extra-virgin olive oil

FOR THE SALAD:

2 cups arugula
2 raw beets, peeled and thinly sliced
1 bulb fennel, shaved
1 red onion, thinly sliced
2 avocados, cubed
½ cup in-season fruit (optional)*
¼ cup shelled pistachios
¼ cup parmesan cheese (optional)

* This is optional—but if you want to add in-season fruit, here are my suggestions:

Early fall: crispy pears
Late fall/winter: pomegranate
Late winter/early spring: grapefruit
Spring: blackberries
Summer: pitted cherries

First make the dressing: Blend all the ingredients in a blender or food processor until smooth.

In a large salad bowl, toss the dressing with the arugula, beets, fennel, and onion until the vegetables are coated. Top with the avocado, fruit, if using, pistachios, and parmesan cheese, if using, and serve.

Sauerkraut Fritters with Leafy Greens

Sauerkraut is one of those foods that's hard for people to get into. It's tangy and a little bitter—and it can be pretty sour. But draining it well and panfrying it until it's crispy makes it the perfect "starter dish" for anyone looking to get into fermented foods. And yes, you can swap out the sauerkraut for cabbage kimchi if you like things a little spicier.

Serves 4.

2 cups well-drained sauerkraut
¼ cup + 1 tablespoon extra-virgin olive oil, divided
1 shallot, minced
1 teaspoon caraway seeds
1 tablespoon fresh rosemary, minced
1 cup leafy greens, minced*
1 large omega-3 or pasture-raised egg
½ cup unsweetened coconut milk
½ cup tapioca starch
½ cup millet flour

Wrap the sauerkraut in a towel and squeeze as much of the moisture out as possible. Spread the sauerkraut onto a dry paper towel and let it continue to air-dry.

In a large pan, heat 1 tablespoon of the oil over medium heat. Add the shallot, caraway seeds, and rosemary and cook until very fragrant, about 1 to 2 minutes. Add the greens, and cook until they are wilted and any liquid in the pan has evaporated, about 5 minutes. Remove from heat and transfer to a bowl. Wipe the pan clean.

In a large bowl, whisk together the egg, coconut milk, tapioca starch, and millet flour to form a thick batter. It should be thicker than pancake batter. Fold in the sauerkraut and cooked greens until a cohesive fritter

*This recipe works with minced kale, chard, dandelion greens, or even spinach.

"dough" is formed. If the mixture feels too wet, sprinkle in additional tapioca starch, 1 tablespoon at a time.

Add the remaining olive oil to the pan over medium heat. When the oil is hot, drop tablespoon-sized balls of batter into the oil (you'll need to work in batches so as not to overcrowd the pan). Cook for 3 to 4 minutes per side, until golden brown and crisp. Remove to a wire rack, and repeat until all dough is used up.

Serve and enjoy.

Goat Milk Yogurt Marinated Curried Chicken Thighs with Basil Sauce

I've always loved using yogurt as a marinade for meat—it adds a delicious tang and really carries the spices in this dish beautifully, for a superflavorful finished product. I've even cooked this chicken on my barbecue grill over a medium-high flame. Be sure to turn it occasionally so it doesn't burn, and cook for about 20 minutes.

Serves 4.

FOR THE CHICKEN:

4 cloves garlic, finely grated or minced
1 tablespoon minced ginger
2 cups unsweetened goat milk yogurt
Juice of 1 lime
1 teaspoon mustard powder
1 teaspoon ground cumin
1 teaspoon iodized sea salt
2 tablespoons curry powder
6 bone-in, skin-on pasture-raised chicken thighs
2 tablespoons extra-virgin olive oil

FOR THE SAUCE:

1 cup fresh basil leaves
2 cloves garlic
2 tablespoons fresh mint
Juice of 1 lime
½ teaspoon Dijon mustard
1 ripe avocado

½ teaspoon sea salt
2 tablespoons apple cider vinegar
Salt and pepper to taste

In a large bowl, combine the garlic, ginger, goat milk yogurt, lime juice, mustard powder, cumin, sea salt, and curry powder. Add the chicken, making sure that it is thoroughly coated with the marinade. Refrigerate for at least four hours or overnight.

In a blender, combine the basil, garlic, mint, lime juice, mustard, avocado, and sea salt. With the blender running, stream in the vinegar until well combined, smooth, and creamy. Taste and adjust the seasoning as necessary (if it is too tangy, add a bit of olive oil to counterbalance the acid). Add salt and pepper to taste.

Remove the chicken from the marinade, and pat dry with paper towels.

Preheat your oven to 400°F. Heat an oven-safe skillet over medium-high heat. When the skillet is hot, add the olive oil. Cook the chicken skin side down for 5 to 7 minutes, until the skin is crispy. Flip the chicken and remove from the stove. Transfer to the oven and bake for 25 to 35 minutes uncovered, until a thermometer inserted in the meatiest part of the chicken reads 160°F.

Serve with the sauce.

Cod and Kimchi Stew

Kimchi-jjigae is the classic Korean kimchi stew, which is often made with kimchi, tofu, and pork. This variation uses the rich, savory flavor of both kimchi and miso, along with gently cooked cod or other whitefish for a really satisfying meal. I know that some people aren't a fan of the texture of kimchi but love the taste. If you're in that category, try radish kimchi instead of a cabbage-based one for a crunchier dish.

Serves 4.

3 tablespoons sesame oil
¼ pound (4 ounces) shiitake or cremini mushrooms, cleaned and halved
1 bulb fennel, minced
3 medium shallots, diced
¼ cup finely minced ginger
5 cloves garlic, minced
1 tablespoon gochujang (Korean fermented pepper paste)
1 tablespoon white miso paste
2 cups kimchi with its juice*
¼ cup coconut aminos
1 pound boneless, skinless cod, halibut, or Lake Superior whitefish, cubed
¼ cup sliced scallions, for serving
1 diced avocado, for serving
Toasted sesame seeds, for serving

Heat the oil in a soup pot over medium-high heat. Sauté the mushrooms and fennel until lightly browned, about 7 to 8 minutes, stirring occasionally. Reduce the heat to medium, and add the shallots, ginger, and garlic. Cook for 4 to 5 minutes, until the ginger is lightly browned and

*If your kimchi has very large pieces of cabbage, I suggest cutting them into bite-sized chunks before starting this recipe.

the garlic is very fragrant. Stir in the gochujang and miso paste and cook until rich in color, about 1 minute.

Add the kimchi, using the juices to deglaze the pan. Scrape any cooked-on bits from the bottom of your pan and cook until most of the liquid is evaporated.

Add the coconut aminos and 5 cups of water and bring to a boil. Reduce the heat to low, then cook uncovered for about 20 minutes, until the kimchi is tender.*

Add the fish and cook on low heat for 2 to 3 minutes, then remove from the heat and cover. Let sit for 5 minutes, while the heat from the broth delicately cooks the fish.

Divide the stew into bowls, top with scallions, avocado, and sesame seeds, and serve.

*If you use radish kimchi, it will retain a crunch even after 20 minutes of cooking. That's totally okay!

Tuna Salad with Sesame Oil and Sauerkraut

If you prefer a vegetarian option, you can swap the tuna for canned hearts of palm or even pressure-cooked chickpeas. You can certainly make a sandwich with this (using lectin-free bread), but it's also great with veggies or as a breakfast along with pasture-raised eggs.

Serves 2.

½ cup sauerkraut, drained
1 can wild-caught tuna in water, drained
1 shallot, minced
1 scallion, finely chopped
1 tablespoon tahini
1½ teaspoons toasted sesame oil
1 teaspoon Dijon mustard
1 ripe avocado, mashed
Juice of ½ lemon
Iodized sea salt to taste

Wrap the sauerkraut in a clean kitchen towel and squeeze as much moisture from it as possible. Transfer the sauerkraut to a bowl and fold in the tuna, shallot, and scallion.

In a large bowl, whisk together the tahini, sesame oil, and mustard until combined. Add the avocado and lemon juice and whisk/mash until it is the consistency of an extra-thick mayo. Fold the "mayo" into the tuna mixture. Taste and add salt as needed.

Serve on lectin-free bread or crackers, or alongside a green salad.

Miso "Grain" Salad with Broccoli and Wild Shrimp

I've always enjoyed a hearty grain salad, such as a rice salad or tabbouleh, but they can be loaded with lectins! This "grain" bowl relies on lectin-free seeds, protein-rich hemp hearts, and tons of herbs for the base. You could even leave out the broccoli and shrimp and just use this as a side dish for one of your other favorite meals.

Serves 4.

Miso Caesar Dressing (page 209)
2 cups cooked millet or sorghum
1 cup hemp hearts
¼ cup ground flaxseeds
¼ cup minced parsley
¼ cup minced mint
¼ cup minced basil
Juice of 1 lemon
¼ cup olive oil, divided
2 cups broccoli florets
2 cups wild-caught shrimp, shells off
3 cloves garlic, minced
1½ teaspoons iodized sea salt, or to taste
¼ cup toasted walnuts
1 cup radish kimchi (optional)

Prepare the Miso Caesar Dressing (instructions are on page 209).

In a large bowl, toss together the millet or sorghum, hemp hearts, flaxseeds, parsley, mint, basil, lemon juice, and dressing. Set aside for at least 10 minutes or as long as 2 hours for the flavors to meld.

Heat a large skillet over medium-high heat. Add half the olive oil and the broccoli. Cook, stirring frequently, until the broccoli is tender and a little charred at the edges, 7 to 10 minutes.

Transfer the broccoli to a separate bowl, and return the pan to the heat.

Add the remaining oil and the shrimp, and cook for 2 to 3 minutes. Flip the shrimp, add the garlic, and cook for an additional 2 to 3 minutes, until the shrimp are cooked through and the garlic is fragrant.

Taste the salad and add salt as needed—it's important to do this *after* the flavors have melded to get the true taste of this dish.

Divide the salad into four bowls and serve topped with the broccoli, shrimp, walnuts, and kimchi, if using.

Miso Lamb Burgers with Mint Sauce

Adding red miso paste to these lamb burgers really brings out the meaty, gamey flavor of the lamb, thanks to the nutty sweetness of the miso. I *highly* recommend doubling the sauce recipe and storing extra in your fridge—I've used it as a salad dressing, to drizzle over eggs and avocados at breakfast, even as a dipping sauce for fritters and meat.

Serves 4.

FOR THE BURGERS:

1 pound ground lamb
2 tablespoons red miso paste*
1 pasture-raised egg yolk
Salt and pepper to taste
2 tablespoons olive oil
Goat cheese for serving (optional)

FOR THE SAUCE:

1 cup plain goat or coconut milk yogurt
¼ cup tahini
Juice of ½ lemon
½ teaspoon coconut aminos
¼ cup fresh mint leaves, finely chopped
1 shallot, minced
1 clove garlic, crushed

In a large bowl, combine the lamb, miso paste, and egg yolk and mix well. Add salt and pepper as needed. Let the mixture rest for 5 to 10 minutes while making the sauce.

*Most of the recipes in this book call for white miso paste, as it is the most widely available, and you can use white miso in this recipe as well, but red miso paste really enhances the rich, gamey flavor of the lamb.

In a bowl, whisk together the yogurt, tahini, lemon juice, and coconut aminos until well combined. Fold in the mint, shallot, and garlic. Set aside.

Heat a large skillet over medium-high heat. While the skillet is heating, shape the lamb mixture into four patties. Add the oil to the pan and let heat for 1 minute, then add the lamb patties. Cook on the first side for 3 to 4 minutes, until a crust is formed, then flip and cook for 3 to 4 more minutes, then reduce the heat to low. Continue cooking until the burger is done to your liking (rare: 120 to 125°F; medium rare: 130 to 135°F; medium well done: 150 to 155°F; well done: 160 to 165°F).

Top with the goat cheese, if using, and a dollop of mint sauce and serve.

Lacto-Fermented Overnight "Oats" (hemp + flax + millet)

I know that conventional overnight oats rely on soaking instead of cooking—but in this case, cooking the millet really helps the texture. This quick fermentation process is one of the easiest ways to DIY fermented food. And depending on what you add to the "oats," it can be a great dessert *or* breakfast.

Serves 4.

1½ cups cooked millet
1 cup hemp hearts
½ cup ground flaxseeds
¼ cup plain goat or coconut milk yogurt with live active cultures
2 cups unsweetened coconut milk
Toppings of your choice (see below)

In a large bowl, mix together the millet, hemp hearts, and flaxseeds. Add the yogurt and coconut milk, making sure that the grains are covered by the liquid. Cover the bowl with a tea towel or cheesecloth and leave in a room temperature place in your kitchen for 8 to 12 hours (overnight is perfect). Enjoy as is, or strain off any liquid if you like it thicker.

To make it really special, add your favorite toppings from the list below.

OUR FAVORITE OVERNIGHT "OATS" TOPPINGS COMBOS

- The almond joy: shaved bittersweet chocolate, unsweetened toasted coconut flakes, marcona almonds
- Vacation in a bowl: unsweetened toasted coconut flakes, fresh passion fruit, macadamia nuts

- Forest fruit (only when in season): ¼ cup in-season mixed berries, including blackberries, raspberries, and wild blueberries
- The blueberry pie (only when in season): wild blueberries, a pinch of allspice, orange zest
- The "better than a peanut butter cup": shaved bittersweet chocolate, pistachio butter, chopped pistachios

If you prefer a warm breakfast, it's fine to heat this over low heat before serving it.

Chocolate Goat Milk Yogurt Snack Cake with Pistachio Butter Drizzle

I think the trend of snack cakes—single-layer cakes with a fun drizzle or frosting—is fantastic. It means that you always have a dessert on hand or something to serve guests. And this snack cake bakes so easily, it's the perfect thing to make on a weeknight—or whenever a sugar craving strikes.

Serves 8 to 10.

FOR THE CAKE:

Olive oil spray
2 large pasture-raised eggs
½ cup unsweetened goat milk yogurt
½ cup allulose
1½ teaspoons vanilla extract
1½ teaspoons almond extract
1 cup blanched almond flour
½ cup natural unsweetened cocoa powder
¼ teaspoon iodized sea salt
½ teaspoon baking soda

FOR THE DRIZZLE:

¼ cup pistachio butter
¼ cup unsweetened goat milk yogurt
⅛ cup allulose
¼ teaspoon ground cinnamon
Zest of 1 orange or tangerine

Preheat your oven to 325°F. Grease an eight-inch cake pan with olive oil spray and set aside.

In a large bowl, whisk together the eggs, yogurt, allulose, vanilla extract, and almond extract.

In another bowl, whisk together the almond flour, cocoa powder, salt, and baking soda. Fold the dry ingredients into the wet ingredients and stir until well combined. Pour the batter into the prepared cake pan. Bake at 325°F until a toothpick inserted into the center of the cake comes out clean, 30 to 35 minutes. Remove from the oven and let cool.

While the cake is cooling, make the drizzle: Whisk together the drizzle ingredients and keep stirring until the allulose has melted into the mixture and is no longer granular. Drizzle over the room temperature cake and serve.

Store the leftovers (if there are any!) in the fridge for up to 5 days.

Spiced Frozen Goat Milk Yogurt

The problem with those trendy frozen yogurt shops is how *sugary* everything is, from the yogurt itself to the hard-to-resist candied toppings. This recipe gives you the delicious sweet-and-tangy flavor that makes frozen yogurt so great, but it's designed to love you back, rather than overload your system with sugar.

Serves 4.

½ cup allulose
1 cinnamon stick
1 piece star anise
1 clove
1 strip orange zest
16 ounces unsweetened goat milk yogurt
1 tablespoon barrel-aged whisky or rum (optional)*
Juice of ½ lemon
⅛ teaspoon salt
¼ cup toasted walnuts

If you are using an ice cream maker, make sure it is clean and any core it may need is frozen and ready to churn.

Make the spice syrup: In a small saucepan, heat ½ cup water with the allulose, cinnamon stick, star anise, clove, and orange zest over medium-low heat. Cook for 5 to 10 minutes, stirring occasionally, until the allulose is dissolved. Let cool to room temperature, then strain and refrigerate until cold, about 2 hours.

In a large bowl, mix together the spice syrup, yogurt, whisky, if using, lemon juice, and salt until smooth and creamy. Pour into a running ice

*This is unnecessary for flavor, but the freezing temperature of the alcohol helps keep the yogurt creamy when frozen. Using a barrel-aged spiced rum (as opposed to a standard dark rum) adds a hint of spiciness to the mixture.

cream machine and process according to the manufacturer's instructions until it is the consistency of frozen yogurt. Add the walnuts, process a moment more, then serve immediately or transfer to an airtight container in your freezer.

> If you don't have an ice cream machine, freeze the mixture in a shallow container with a lid, stirring every 30 minutes until it is creamy and frozen.

Fudgy Miso Brownies

Doesn't *everyone* love a rich, fudgy brownie? If you want to really push this recipe over the top, add a scoop of Spiced Frozen Goat Milk Yogurt (page 240) for an irresistible ice cream sundae.

Makes 12 generous-sized brownies.

1 cup powdered (not granulated) allulose or Swerve*
2 cups blanched almond flour
1 cup natural (nondutched) cocoa powder
¾ teaspoon kosher salt
4 egg whites
¼ cup sesame oil
2 tablespoons white miso paste
1½ teaspoons vanilla extract

Preheat your oven to 350°F. Line a 9-by-13-inch baking pan with parchment paper, and set aside.

In a large bowl, mix together the sweetener, almond flour, cocoa powder, and salt. Set aside.

In a separate bowl, whisk together the egg whites, oil, miso paste, and vanilla extract until smooth. You can save time by doing this in a blender or food processor. Fold the wet ingredients into the dry ingredients until you form a thick batter with no dry spots. Pour the batter into the prepared baking pan and bake for 18 to 20 minutes, until a toothpick inserted into the center comes out clean.

Let cool to room temperature, slice, and serve.

*Allulose is my current favorite sweetener, but the powdered version isn't as widely available as powdered (confectioner's) Swerve. Both work for this recipe.

ACKNOWLEDGMENTS

It's always exciting to start writing my next book, because no matter what I set out to write, after falling into multiple "rabbit holes," invariably a more fascinating book comes out on the other side. With that in mind, I invited my collaborator from my *New York Times* bestseller *The Longevity Paradox*, Jodi Lipper, to hop aboard the train again, and, boy, did we enjoy the ride! Jodi, we had so much fun, let's do it again soon! And thanks for making my nerdy, thick, professor-speak science accessible to our readers once again.

The recipes were once again contributed by Kathryn "Kate" Holzhauer, my head chef at GundryMD, aimed at harnessing the power of fermented foods in a tasty way. Kate and I are constantly emailing each other about a fun recipe she or I found that could be tweaked into something delicious that would save your life rather than kill you; and we both think we've succeeded again. We can't wait for you to try them! Thanks again, Kate!

This book is the last hurrah for my longtime (and only) editor at HarperCollins, vice president and editorial director Julie Will, who left for an exciting new opportunity after editing this manuscript. Julie, thanks for such an amazing run of bestsellers, which have affected millions of lives worldwide.

But fear not, the publisher of Harper Wave, Karen Rinaldi, remains at the helm, guiding this book into publication, as well as my longtime editorial assistant, Emma Kupor. Thanks additionally to Kirby Sandmeyer, associate editor; Yelena Nesbit, my longtime executive director of publicity; Amanda Pritzker, senior marketing director; Milan Bozic, my longtime cover designer; and Nancy Singer, interior designer.

Naturally, much of what you read on these pages stems from the fact that I continue to see patients six days a week at the International Heart and Lung Institute in Palm Springs, California, and at the Centers for Restorative Medicine, in Palm Springs, Santa Barbara, and now in Beverly Hills. All of this couldn't happen without the tireless work of Mitsu Killion-Jacobo, my longtime physician's assistant and associate director of the International Heart and Lung Institute; Susan Lokken, my right-hand woman of decades as executive assistant and office manager; Jessenia Parra, my receptionist extraordinaire, who gets everything done; Debbie Stewart, office assistant who arrives smiling to work daily; and, of course, the "blood suckers" Laurie Acuna and Lynn Visk, without whom all the startling results of my program you wouldn't have known about. All this is kept aboveground by my CFO, Joseph Tames, and my friend and attorney, Dave Baron. Thanks to you all!

And thanks to my longtime agent and protector Shannon Marven, president of Dupree Miller, and her assistant and co-worker, Rebecca Silensky.

A continued shout-out to the hundreds of employees at GundryMD who have made me, GundryMD.com, *The Dr. Gundry Podcast*, and my YouTube channels essential sources for your health and wellness needs. And a special shout-out (again) to my support team at GundryMD, headed by Lanee Lee Neil, along with Kate Holzhauer (mentioned above), as well as the talented writers and film crew that keeps the cutting-edge health information that I discover (thanks to my patients) coming out to you.

Despite the fact that I continue to see patients six days a week, even on Saturdays and Sundays, the wait times to see me or Mitsu are very long. Because of that, I've recently launched Gundry Health and its app, GundryHealth.com, my subscription-based telemedicine service, where you can interact with board certified physicians, trained by me, to manage and treat all autoimmune conditions, "leaky gut," and IBS using the same protocol and state-of-the-art blood tests that you read about in this book. Finally, you can now see "me" without the wait! I can't wait to see you there.

Final thanks, of course, goes to my soulmate and wife, Penny, who keeps me grounded and in check while caring for our four dogs, now including two new rescues. She finally retired from her twenty-year-old business, Zense, so has more time to harass me into taking care of myself. Thanks, Penny!

NOTES

INTRODUCTION: BACTERIAL BRAIN WASHING

1. Ehrenberg, R. (2015, September 2). Global forest survey finds trillions of trees. *Nature*. https://doi.org/10.1038/nature.2015.18287
2. Vyas, A., Kim, S. K., Giacomini, N., Boothroyd, J. C., & Sapolsky, R. M. (2007). Behavioral changes induced by *Toxoplasma* infection of rodents are highly specific to aversion of cat odors. *Proceedings of the National Academy of Sciences of the United States of America*, *104*(15), 6442–6447. https://doi.org/10.1073/pnas.0608310104
3. Meyer, C. J., Cassidy, K. A., Stahler, E. E., Brandell, E. E., Anton, C. B., Stahler, D. R., & Smith, D. W. (2022). Parasitic infection increases risk-taking in a social, intermediate host carnivore. *Communications Biology*, *5*, article 1180. https://doi.org/10.1038/s42003-022-04122-0
4. Brandell, E. E., Cross, P. C., Craft, M. E., Smith, D. W., Dubovi, E. J., Gilbertson, M.L.J., Wheeldon, T., Stephenson, J. A., Barber-Meyer, S., Borg, B. L., Sorum, M., Stahler, D. R., Kelly, A., Anderson, M., Cluff, H. D., MacNulty, D. R., Watts, D. E., Roffler, G. H., Schwantje, H., . . . Hudson, P. J. (2021). Patterns and processes of pathogen exposure in gray wolves across North America. *Scientific Reports*, *11*, article 3722. https://doi.org/10.1038/s41598-021-81192-w
5. Flegr, J. (2013). Influence of latent *Toxoplasma* infection on human personality, physiology and morphology: pros and cons of the *Toxoplasma*-human model in studying the manipulation hypothesis. *Journal of Experimental Biology*, *216*(1), 127–133. https://doi.org/https://doi.org/10.1242/jeb.073635
6. Flegr, J., Havlicek, J., Kodym, P., Malý, M., & Smahel, Z. (2002). Increased risk of traffic accidents in subjects with latent toxoplasmosis: A retrospective case-control study. *BMC Infectious Diseases*, *2*, article 11. https://www.ncbi.nlm.nih.gov/pmc/articles/PMC117239/#!po=70.8333
7. Virus, M. A., Ehrhorn, E. G., Lui, L. M., & Davis, P. H. (2021). Neurological and neurobehavioral disorders associated with *Toxoplasma gondii* infection in humans. *Journal of Parasitology Research*, article 6634807. https://doi.org/10.1155/2021/6634807

8. Poirotte, C., Kappeler, P., Ngoubangoye, B., Bourgeois, S., Moussodji, M., & Charpentie, M. (2016). Morbid attraction to leopard urine in *Toxoplasma*-infected chimpanzees. *Current Biology*, *26*, R98–R99. https://doi.org/10.1016/j.cub.2015.12.020

CHAPTER 1: YOUR BODY IS A RAIN FOREST

1. Lozupone, C. A., Stombaugh, J. I., Gordon, J. I., Jansson, J. K., & Knight, R. (2012). Diversity, stability and resilience of the human gut microbiota. *Nature*, *489*(7415), 220–230. https://doi.org/10.1038/nature11550
2. Hatch, M., & Freel, R. W. (2008). The roles and mechanisms of intestinal oxalate transport in oxalate homeostasis. *Seminars in Nephrology*, *28*(2), 143–151. https://doi.org/10.1016/j.semnephrol.2008.01.007
3. Daisley, B. A., Koenig, D., Engelbrecht, K., Doney, L., Hards, K., Al, K. F., Reid, G., & Burton, J. P. (2021). Emerging connections between gut microbiome bioenergetics and chronic metabolic diseases, *Cell Reports*, *37*(10), article 110087. https://doi.org/10.1016/j.celrep.2021.110087
4. Annunziata, G., Maisto, M., Schisano, C., Ciampaglia, R., Narciso, V., Tenore, G. C., & Novellino, E. (2019). Effects of grape pomace polyphenolic extract (Taurisolo®) in reducing TMAO serum levels in humans: Preliminary results from a randomized, placebo-controlled, cross-over study. *Nutrients*, *11*(1), 139. https://doi.org/10.3390/nu11010139
5. Relman, D. A. (2012). The human microbiome: ecosystem resilience and health. *Nutrition Reviews*, *70*(Suppl. 1), S2–S9. https://doi.org/10.1111/j.1753-4887.2012.00489.x
6. Lu, K., Cable, P. H., Abo, R. P., Ru, H., Graffam, M. E., Schlieper, K. A., Parry, N. M., Levine, S., Bodnar, W. M., Wishnok, J. S., Styblo, M., Swenberg, J. A., Fox, J. G., & Tannenbaum, S. R. (2013). Gut microbiome perturbations induced by bacterial infection affect arsenic biotransformation. *Chemical Research in Toxicology*, *26*(12), 1893–1903. https://doi.org/10.1021/tx4002868
7. Dethlefsen, L., & Relman, D. A. (2010). Incomplete recovery and individualized responses of the human distal gut microbiota to repeated antibiotic perturbation. *Proceedings of the National Academy of Sciences of the United States of America*, *108*(Suppl. 1), 4554–4561. https://doi.org/10.1073/pnas.1000087107
8. Relman, D. A. (2012). The human microbiome: ecosystem resilience and health. *Nutrition Reviews*, *70*(Suppl. 1), S2–S9. https://doi.org/10.1111/j.1753-4887.2012.00489.x
9. Lozupone, C. A., Stombaugh, J. I., Gordon, J. I., Jansson, J. K., & Knight, R. (2012). Diversity, stability and resilience of the human gut microbiota. *Nature*, *489*(7415), 220–230. https://doi.org/10.1038/nature11550
10. Hsiao, E. Y., McBride, S. W., Hsien, S., Sharon, G., Hyde, E. R., McCue, T., Codelli, J. A., Chow, J., Reisman, S. E., Petrosino, J. F., Patterson, P. H., & Mazmanian, S. K. (2013). Microbiota modulate behavioral and physiological abnormalities associated with neurodevelopmental disorders. *Cell*, *155*(7), 1451–1463. https://doi.org/10.1016/j.cell.2013.11.024
11. Mazmanian, S. K., Round, J. L., & Kasper, D. L. (2008). A microbial symbiosis

factor prevents intestinal inflammatory disease. *Nature*, *453*(7195), 620–625. https://doi.org/10.1038/nature07008

12. Lozupone, C. A., Stombaugh, J. I., Gordon, J. I., Jansson, J. K., & Knight, R. (2012). Diversity, stability and resilience of the human gut microbiota. *Nature*, *489*(7415), 220–230. https://doi.org/10.1038/nature11550
13. Lee, Y. K., & Mazmanian, S. K. (2010). Has the microbiota played a critical role in the evolution of the adaptive immune system? *Science*, *330*(6012), 1768–1773. https://doi.org/10.1126/science.1195568
14. Eberl, G. (2010). A new vision of immunity: Homeostasis of the superorganism. *Mucosal Immunology*, *3*(5), 450–460.
15. Bever, J. D., Westover, K. M., & Antonovics, J. (1997). Incorporating the soil community into plant population dynamics: The utility of the feedback approach. *Journal of Ecology*, *85*(5), 561–573.
16. Banerjee, S., Schlaeppi K., & van der Heijden, M.G.A. (2018). Keystone taxa as drivers of microbiome structure and functioning. *Nature Reviews Microbiology*, *16*(9), 567–576. https://doi.org/10.1038/s41579-018-0024-1
17. Bäckhed, F., Ley, R. E., Sonnenburg, J. L., Peterson, D. A., & Gordon, J. I. (2005). Host-bacterial mutualism in the human intestine. *Science*, *307*(5717), 1915–1920. https://doi.org/10.1126/science.1104816
18. Kong, F., Hua, Y., Zeng, B., Ning, R., Li, Y., & Zhao, J. (2016). Gut microbiota signatures of longevity. *Current Biology*, *26*(18), R832–R833. https://doi.org/10.1016/j.cub.2016.08.015
19. David, L. A., Weil, A., Ryan, E. T., Calderwood, S. B., Harris, J. B., Chowdhury, F., Begum, Y., Qadri, F., LaRocque, R. C., & Turnbaugh, P. J. (2015). Gut microbial succession follows acute secretory diarrhea in humans. *MBio*, *6*(3), e00381–e00315.
20. Lozupone, C. A., Stombaugh, J., Gonzalez, A., Ackermann, G., Wendel, D., Vazquez-Baeza, Y., Jansson, J. K., Gordon, J. I., & Knight, R. (2013). Meta-analyses of studies of the human microbiota. *Genome Research*, *23*, 1704–1714.
21. Qin, N., Yang, F., Li, A., Prifti, E., Chen, Y., Shao, L., Guo, J., Le Chatelier, E., Yao, J., Wu, L., Zhou, J., Ni, S., Liu, L., Pons, N., Batto, J. M., Kennedy, S. P., Leonard, P., Yuan, C., Ding, W., . . . Li, L. (2014). Alterations of the human gut microbiome in liver cirrhosis. *Nature*, *513*(7516), 59–64.
22. Liu, C., Frank, D. N., Horch, M., Chau, S., Ir, D., Horch, E. A., Tretina, K., van Besien, K., Lozupone, C. A., & Nguyen, V. H. (2017). Associations between acute gastrointestinal GvHD and the baseline gut microbiota of allogeneic hematopoietic stem cell transplant recipients and donors. *Bone Marrow Transplantation*, *52*(12), 1643–1650.
23. Menni, C., Jackson, M. A., Pallister, T., Steves, C. J., Spector, T. D., & Valdes, A. M. (2017). Gut microbiome diversity and high-fibre intake are related to lower long-term weight gain. *International Journal of Obesity*, *41*(7), 1099–1105. https://doi.org/10.1038/ijo.2017.66
24. Lozupone, C. A., Stombaugh, J. I., Gordon, J. I., Jansson, J. K., & Knight, R. (2012). Diversity, stability and resilience of the human gut microbiota. *Nature*, *489*(7415), 220–230. https://doi.org/10.1038/nature11550
25. Lozupone, C., Faust, K., Raes, J., Faith, J. J., Frank, D. N., Zaneveld, J., Gordon,

J. I., & Knight R. (2012). Identifying genomic and metabolic features that can underlie early successional and opportunistic lifestyles of human gut symbionts. *Genome Research*, *22*(10), 1974–1984.
26. Gutiérrez, N., & Garrido, D. (2019). Species deletions from microbiome consortia reveal key metabolic interactions between gut microbes. *mSystems*, *4*(4), e00185-19. https://doi.org/10.1128/mSystems.00185-19
27. Corrêa-Oliveira, R., Fachi, J. L., Vieira, A., Sato, F. T., & Vinolo, M. A. (2016). Regulation of immune cell function by short-chain fatty acids. *Clinical & Translational Immunology*, *5*(4), e73. https://doi.org/10.1038/cti.2016.17
28. Parfrey, L. W., Walters, W. A., & Knight, R. (2011). Microbial eukaryotes in the human microbiome: ecology, evolution, and future directions. *Frontiers in Microbiology*, *2*, 153. https://doi.org/10.3389/fmicb.2011.00153
29. Lukeš, J., Kuchta, R., Scholz, T., & Pomajbíková, K. (2014). (Self-) infections with parasites: Re-interpretations for the present. *Trends in Parasitology*, *30*(8), 377–385. https://doi.org/10.1016/j.pt.2014.06.005
30. Foster, K. R., & Bell, T. (2012). Competition, not cooperation, dominates interactions among culturable microbial species. *Current Biology*, *22*(19), 1845–1850.
31. Kim, H. J., Boedicker, J. Q., Choi, J. W., & Ismagilov, R. F. (2008). Defined spatial structure stabilizes a synthetic multispecies bacterial community. *Proceedings of the National Academy of Sciences of the United States of America*, *105*(47), 18188–18193.
32. Willcox, M. D., Zhu, H., Conibear, T. C., Hume, E. B., Givskov, M., Kjelleberg, S., & Rice, S. A. (2008). Role of quorum sensing by *Pseudomonas aeruginosa* in microbial keratitis and cystic fibrosis. *Microbiology*, *154*(part 8), 2184–2194.
33. Waters, C. M., & Bassler, B. L. (2005). Quorum sensing: Cell-to-cell communication in bacteria. *Annual Review of Cell and Developmental Biology*, *21*, 319–346. https://doi.org/10.1146/annurev.cellbio.21.012704.131001
34. Fiegna, F., & Velicer, G. J. (2005). Exploitative and hierarchical antagonism in a cooperative bacterium. *PLOS Biology*, *3*(11), e370. https://doi.org/10.1371/journal.pbio.0030370
35. Fiegna, F., & Velicer, G. J. (2005). Exploitative and hierarchical antagonism in a cooperative bacterium. *PLOS Biology*, *3*(11), e370. https://doi.org/10.1371/journal.pbio.0030370

CHAPTER 2: IT TAKES TWO

1. Samczuk, P., Hady, H. R., Adamska-Patruno, E., Citko, A., Dadan, J., Barbas, C., Kretowski, A., & Ciborowski, M. (2018). In-and-out molecular changes linked to the type 2 diabetes remission after bariatric surgery: An influence of gut microbes on mitochondria metabolism. *International Journal of Molecular Sciences*, *19*(12), 3744. https://doi.org/10.3390/ijms19123744
2. Speakman, J. R., Talbot, D. A., Selman, C., Snart, S., McLaren, J. S., Redman, P., Krol, E., Jackson, D. M., Johnson, M. S., & Brand, M. D. (2004). Uncoupled and surviving: Individual mice with high metabolism have greater mitochon-

drial uncoupling and live longer. *Aging Cell*, *3*(3), 87–95. https://doi.org/10.1111/j.1474-9728.2004.00097.x

3. Cortés-Martín, A., Selma, M. V., Tomás-Barberán, F. A., González-Sarrías, A., & Espín, J. C. (2020). Where to look into the puzzle of polyphenols and health? The postbiotics and gut microbiota associated with human metabotypes. *Molecular Nutrition & Food Research*, *64*(9), article e1900952. https://doi.org/10.1002/mnfr.201900952
4. Brand, M. (2000). Uncoupling to survive? The role of mitochondrial inefficiency in ageing. *Experimental Gerontology*, *35*(6–7), 811–820. https://doi.org/10.1016/s0531-5565(00)0013
5. Chandrasekaran, K., Salimian, M., Konduru, S. R., Choi, J., Kumar, P., & Long, A. (2019). Overexpression of Sirtuin 1 protein in neurons prevents and reverses experimental diabetic neuropathy. *Brain*, *142*(12), 3737–3752.
6. Vauzour, D., Houseman, E., George, T., Corona, G., Garnotel, R., Jackson, K., Sellier, C., Gillery, P., Kennedy, O., Lovegrove, J., & Spencer, J. (2010). Moderate Champagne consumption promotes an acute improvement in acute endothelial-independent vascular function in healthy human volunteers. *British Journal of Nutrition*, *103*(8), 1168–1178. https://doi.org/10.1017/S0007114509992959
7. Ma, X., Sun, Z., Han, X., Li, S., Jiang, X., & Chen, S. (2019). Neuroprotective effect of resveratrol via activation of Sirt1 signaling in a rat model of combined diabetes and Alzheimer's disease. *Frontiers in Neuroscience*, *13*, 1400.
8. Dasgupta, B., & Milbrandt, J. (2007). Resveratrol stimulates AMP kinase activity in neurons. *Proceedings of the National Academy of Sciences of the United States of America*, *104*(17), 7217–7222.
9. Rege, S. D., Geetha, T., Griffin, G. D., Broderick, T. L., & Babu, J. R. (2014). Neuroprotective effects of resveratrol in Alzheimer disease pathology. *Frontiers in Aging Neuroscience*, *6*, 218.
10. Ryu, D., Mouchiroud, L., Andreux, P. A., Katsyuba, E., Moullan, N., Nicolet-dit-Félix, A. A., Williams, E. G., Jha, P., Lo Sasso, G., Huzard, D., Aebischer, P., Sandi, C., Rinsch, C., & Auwerx, J. (2016). Urolithin A induces mitophagy and prolongs lifespan in *C. elegans* and increases muscle function in rodents. *Nature Medicine*, *22*(8), 879–888. https://doi.org/10.1038/nm.4132
11. D'Amico, D., Olmer, M., Fouassier, A. M., Valdés, P., Andreux, P. A., Rinsch, C., & Lotz, M. (2022). Urolithin A improves mitochondrial health, reduces cartilage degeneration, and alleviates pain in osteoarthritis. *Aging Cell*, *21*(8), article e13662. https://doi.org/10.1111/acel.13662
12. Selma, M. V., Beltrán, D., Luna, M. C., Romo-Vaquero, M., García-Villalba, R., Mira, A., Espín, J. C., & Tomás-Barberán, F. A. (2017). Isolation of human intestinal bacteria capable of producing the bioactive metabolite isourolithin A from ellagic acid. *Frontiers in Microbiology*, *8*, 1521. https://doi.org/10.3389/fmicb.2017.01521
13. Schönfeld, P., Wojtczak, A. B., Geelen, M.J.H., Kunz, W., & Wojtczak, L. (1988). On the mechanism of the so-called uncoupling effect of medium- and short-chain fatty acids. *Biochimica et Biophysica Acta—Bioenergetics*, *936*(3), 280–288.

14. Wastyk, H. C., Fragiadakis, G. K., Perelman, D., Dahan, D., Merrill, B. D., Yu, F. B., Topf, M., Gonzalez, C. G., Van Treuren, W., Han, S., Robinson, J. L., Elias, J. E., Sonnenberg, E. D., Gardner, C. D., & Sonnenburg, J. L. (2021). Gut-microbiota-targeted diets modulate human immune status. *Cell*, *184*(16), 4137–4153.e14. https://doi.org/10.1016/j.cell.2021.06.019
15. Duncan, S. H., Louis, P., & Flint, H. J. (2004). Lactate-utilizing bacteria, isolated from human feces, that produce butyrate as a major fermentation product. *Applied and Environmental Microbiology*, *70*(10), 5810–5817.
16. Belenguer, A., Duncan, S. H., Calder, A. G., Holtrop, G., Louis, P., Lobley, G. E., & Flint, H. J. (2006). Two routes of metabolic cross-feeding between *Bifidobacterium adolescentis* and butyrate-producing anaerobes from the human gut. *Applied and Environmental Microbiology*, *72*(5), 3593–3599.
17. Li, Z., Teng, J., Lyu, Y., Hu, X., Zhao, Y., & Wang, M. (2018). Enhanced antioxidant activity for apple juice fermented with *Lactobacillus plantarum* ATCC14917. *Molecules*, *24*(1), 51. https://doi.org/10.3390/molecules24010051
18. Chia, L. W., Hornung, B.V.H., Aalvink, S., Schaap, P. J., De Vos, W. M., Knol, J., & Belzer, C. (2018). Deciphering the trophic interaction between *Akkermansia muciniphila* and the butyrogenic gut commensal *Anaerostipes caccae* using a metatranscriptomic approach. *Antonie van Leeuwenhoek*, *111*(6), 859–873.
19. Burger-van Paassen, N., Vincent, A., Puiman, P. J., van der Sluis, M., Bouma, J., Boehm, G., van Goudoever, J. B., van Seuningen, I., & Renes, I. B. (2009). The regulation of intestinal mucin MUC2 expression by short-chain fatty acids: implications for epithelial protection. *Biochemical Journal*, *420*(2), 211–219.
20. Gaudier, E., Rival, M., Buisine, M. P., Robineau, I., & Hoebler, C. (2009). Butyrate enemas upregulate muc genes expression but decrease adherent mucus thickness in mice colon. *Physiological Research*, *58*(1), 111–119.
21. Ferreira, T. M., Leonel, A. J., Melo, M. A., Santos, R. R., Cara, D. C., Cardoso, V. N., Correia, M. I., & Alvarez-Leite, J. I. (2012). Oral supplementation of butyrate reduces mucositis and intestinal permeability associated with 5-fluorouracil administration. *Lipids*, *47*(7), 669–678.
22. Bjursell, M., Admyre, T., Göransson, M., Marley, A. E., Smith, D. M., Oscarsson, J., & Bohlooly-Y, M. (2011). Improved glucose control and reduced body fat mass in free fatty acid receptor 2–deficient mice fed a high-fat diet. *American Journal of Physiology: Endocrinology and Metabolism*, *300*(1), E211–E220.
23. Frost, G., Sleeth, M. L., Sahuri-Arisoylu, M., Lizarbe, B., Cerdan, S., Brody, L., Anastasovska, J., Ghourab, S., Hankir, M., Zhang, S., Carling, D., Swann, J. R., Gibson, G., Viardot, A., Morrison, D., Thomas, E. L., & Bell, J. D. (2014). The short-chain fatty acid acetate reduces appetite via a central homeostatic mechanism. *Nature Communications*, *5*, article 3611.
24. Cani, P. D., Lecourt, E., Dewulf, E. M., Sohet, F. M., Pachikian, B. D., Naslain, D., De Backer, F., Neyrinck, A. M., & Delzenne, N. M. (2009). Gut microbiota fermentation of prebiotics increases satietogenic and incretin gut peptide production with consequences for appetite sensation and glucose response after a meal. *The American Journal of Clinical Nutrition*, *90*(5), 1236–1243.
25. Bain, M. D., Borriello, S. P., Tracey, B. M., Jones, M., Reed, P. J., Chalmers, R.

A., & Stacey, T. E. (1988). Contribution of gut bacterial metabolism to human metabolic disease. *The Lancet*, *1*(8594), 1078–1079.

26. Masui, R., Sasaki, M., Funaki, Y., Ogasawara, N., Mizuno, M., Iida, A., Izawa, S., Kondo, Y., Ito, Y., Tamura, Y., Yanamoto, K., Noda, H., Tanabe, A., Okaniwa, N., Yamaguchi, Y., Iwamoto, T., & Kasugai, K. (2013). G protein–coupled receptor 43 moderates gut inflammation through cytokine regulation from mononuclear cells. *Inflammatory Bowel Diseases*, *19*(13), 2848–2856.
27. Hamer, H. M., Jonkers, D.M.A.E., Bast, A., Vanhoutvin, S.A.L.W., Fischer, M.A.J.G., Kodde, A., Troost, F. J., Venema, K., & Brummer, R.-J. M. (2009). Butyrate modulates oxidative stress in the colonic mucosa of healthy humans. *Clinical Nutrition*, *28*(1), 88–93.
28. Kaisar, M.M.M., Pelgrom, L. R., Van der Ham, A. J., Yazdanbakhsh, M., & Everts, B. (2017). Butyrate conditions human dendritic cells to prime type 1 regulatory T cells via both histone deacetylase inhibition and G protein–coupled receptor 109A signaling. *Frontiers in Immunology*, *8*, 1429.
29. Goverse, G., Molenaar, R., Macia, L., Tan, J., Erkelens, M. N., Konijn, T., Knippenberg, M., Cook, E.C.L., Hanekamp, D., Veldhoen, M., Hartog, A., Roeselers, G., Mackay, C. R., & Mebius, R. E. (2017). Diet-derived short chain fatty acids stimulate intestinal epithelial cells to induce mucosal tolerogenic dendritic cells. *Journal of Immunology*, *198*(5), 2172–2181.
30. Arpaia, N., Campbell, C., Fan, X., Dikiy, S., van der Veeken, J., deRoos, P., Liu, H., Cross, J. R., Pfeffer, K., Coffer, P. J., & Rudensky, A. Y. (2013). Metabolites produced by commensal bacteria promote peripheral regulatory T-cell generation. *Nature*, *504*(7480), 451–455.
31. Smith, K., McCoy, K. D., & Macpherson, A. J. (2007). Use of axenic animals in studying the adaptation of mammals to their commensal intestinal microbiota. *Seminars in Immunology*, *19*(2), 59–69.
32. Pabst, O., & Slack, E. (2020). IgA and the intestinal microbiota: The importance of being specific. *Mucosal Immunology*, *13*(1), 12–21. https://doi.org/10.1038/s41385-019-0227-4
33. Norgren, J., Sindi, S., Sandebring-Matton, A., Kåreholt, I., Daniilidou, M., Akenine, U., Nordin, K., Rosenborg, S., Ngandu, T., & Kivipelto, M. (2020). Ketosis after intake of coconut oil and caprylic acid—with and without glucose: A cross-over study in healthy older adults. *Frontiers in Nutrition*, *7*, 40. https://www.ncbi.nlm.nih.gov/pmc/articles/PMC7175812/
34. Kaviyarasan, S., Chung Sia, E. L., Retinasamy, T., Arulsamy, A., & Shaikh, M. F. (2022). Regulation of gut microbiome by ketogenic diet in neurodegenerative diseases: A molecular crosstalk. *Frontiers in Aging Neuroscience*, *14*, article 1015837. https://doi.org/10.3389/fnagi.2022.1015837
35. Castosa, R., Martinez-Iglesias, O., Roca-Lema, D., Casas-Pais, A., Diaz-Diaz, A., Iglesias, P., Santamarina, I., Graña, B., Cavo, L., Valladares-Ayerbes, M., Concha, Á., & Figueroa, A. (2018). Hakai overexpression effectively induces tumour progression and metastasis *in vivo*. *Scientific Reports*, *8*(1), article 3466. https://doi.org/10.1038/s41598-018-21808-w
36. Chen, X.-M., Huang, Q.-C., Yang, S.-L., Chu, Y.-L., Yan, Y.-H., Han, L., Huang, Y., & Huang, R.-Y. (2015). Role of micro RNAs in the pathogenesis

of rheumatoid arthritis: Novel perspectives based on review of the literature. *Medicine*, *94*(31), e126. https://doi.org/10.1097/md.0000000000001326

37. Chen, J., Papp, G., Szodoray, P., & Zeher, M. J. (2016). The role of microRNAs in the pathogenesis of autoimmune diseases. *Autoimmunity Reviews*, *15*(12), 1171–1180. https://doi.org/10.1016/j.autrev.2016.09.003
38. Liu, S., Pires da Cunha, A., Rezende, R. M., Cialic, R., Wei, Z., Bry, L., Comstock, L. E., Gandhi, R., & Weiner, H. L. (2016). The host shapes the gut microbiota via fecal microRNA. *Cell Host & Microbe*, *19*(1), 32–43. https://doi.org/10.1016/j.chom.2015.12.005
39. Bartel, D. P. (2009). MicroRNAs: Target recognition and regulatory functions. *Cell*, *136*(2), 215–233.
40. Hogan, S. P., Seidu, L., Blanchard, C., Groschwitz, K., Mishra, A., Karow, M. L., Ahrens, R., Artis, D., Murphy, A. J., Valenzuela, D. M., Yancopolous, G. D., & Rothenberg, M. E. (2006). Resistin-like molecule beta regulates innate colonic function: Barrier integrity and inflammation susceptibility. *Journal of Allergy and Clinical Immunology*, *118*(1), 257–268.
41. Ahmed, F. E., Jeffries, C. D., Vos, P. W., Flake, G., Nuovo, G. J., Sinar, D. R., Naziri, W., & Marcuard, S. P. (2009). Diagnostic microRNA markers for screening sporadic human colon cancer and active ulcerative colitis in stool and tissue. *Cancer Genomics & Proteomics*, *6*(5), 281–295.
42. Kumar, M., Lu, J., Mercer, K., Golub, T., & Jacks, T. (2007). Impaired microRNA processing enhances cellular transformation and tumorigenesis. *Nature Genetics*, *39*(5), 673–677. https://doi.org/10.1038/ng2003
43. Hu, S., Liu, L., Chang, E. B., Wang, J. Y., & Raufman, J. P. (2015). Butyrate inhibits pro-proliferative miR-92a by diminishing c-Myc-induced miR-17-92a cluster transcription in human colon cancer cells. *Molecular Cancer*, *14*, 180. https://doi.org/10.1186/s12943-015-0450-x
44. Liu, S., Pires da Cunha, A., Rezende, R. M., Cialic, R., Wei, Z., Bry, L., Comstock, L. E., Gandhi, R., & Weiner, H. L. (2016). The host shapes the gut microbiota via fecal microRNA. *Cell Host & Microbe*, *19*(1), 32–43. https://doi.org/10.1016/j.chom.2015.12.005
45. Shemarova, I., Nesterov, V., Emelyanova, L., & Korotkov, S. (2021). Mitochondrial mechanisms by which gasotransmitters (H2S, NO and CO) protect cardiovascular system against hypoxia. *Frontiers in Bioscience—Scholar*, *13*(2), 105–130.
46. Woller, S. A., Eddinger, K. A., Corr, M., & Yaksh, T. L. (2017). An overview of pathways encoding nociception. *Clinical and Experimental Rheumatology*, *35*(5) (Suppl. 107), 40–46.
47. Zhang, W., Lyu, M., Bessman, N. J., Xie, Z., Arifuzzaman, M., Yano, H., Parkhurst, C. N., Chu, C., Zhou, L., Putzel, G. G., Li, T. T., Jin, W. B., Zhou, J., JRI Live Cell Bank, Hu, H., Tsou, A. M., Guo, C. J., & Artis, D. (2022). Gut-innervating nociceptors regulate the intestinal microbiota to promote tissue protection. *Cell*, *185*(22), 4170–4189.e20. https://doi.org/10.1016/j.cell.2022.09.008
48. Li, Y.-L., Wu, P.-F., Chen, J.-G., Wang, S., Han, Q.-Q., Li, D., Wang, W., Guan, X.-L., Li, D., Long, L.-H., Huang, J.-G., & Wang, F. (2017). Activity-

dependent sulfhydration signal controls N-methyl-D-aspartate subtype glutamate receptor-dependent synaptic plasticity *via* increasing d-serine availability. *Antioxidants & Redox Signaling*, *27*(7), 398–414.

49. Hou, X.-Y., Hu, Z.-L., Zhang, D.-Z., Lu, W., Zhou, J., Wu, P.-F., Guan, X.-L., Han, Q.-Q., Deng, S.-L., Zhang, H., Chen, J.-G., & Wang, F. (2017). Rapid antidepressant effect of hydrogen sulfide: evidence for activation of mTORC1-TrkB-AMPA receptor pathways. *Antioxidants & Redox Signaling*, *27*(8), 472–488.
50. Szabó, C., & Papapetropoulos, A. (2011). Hydrogen sulphide and angiogenesis: Mechanisms and applications. *British Journal of Pharmacology*, *164*(3), 853–865. https://doi.org/10.1111/j.1476-5381.2010.01191.x
51. Motta, J. P., Flannigan, K. L., Agbor, T. A., Beatty, J. K., Blackler, R. W., Workentine, M. L., Da Silva, G. J., Wang, R., Buret, A. G., & Wallace, J. L. (2015). Hydrogen sulfide protects from colitis and restores intestinal microbiota biofilm and mucus production. *Inflammatory Bowel Diseases*, *21*(5), 1006–1017.
52. Szabo, C., & Papapetropoulos, A. (2017). International Union of Basic and Clinical Pharmacology. CII: Pharmacological modulation of H_2S levels: H_2S donors and H_2S biosynthesis inhibitors. *Pharmacological Reviews*, *69*(4), 497–564.
53. Hine, C., Harputlugil, E., Zhang, Y., Ruckenstuhl, C., Lee, B. C., Brace, L., Longchamp, A., Treviño-Villarreal, J. H., Mejia, P., Ozaki, C. K., Wang, R., Gladyshev, V. N., Madeo, F., Mair, W. B., & Mitchell, J. R. (2015). Endogenous hydrogen sulfide production is essential for dietary restriction benefits. *Cell*, *160*(1–2), 132–144.
54. Buret, A. G., Allain, T., Motta, J. P., & Wallace, J. L. (2022). Effects of hydrogen sulfide on the microbiome: From toxicity to therapy. *Antioxidants & Redox Signaling*, *36*(4–6), 211–219. https://doi.org/10.1089/ars.2021.0004
55. Kolluru, G. K., Shen, X., Bir, S. C., & Kevil, C. G. (2013). Hydrogen sulfide chemical biology: Pathophysiological roles and detection. *Nitric Oxide*, *35*, 5–20.
56. Kolluru, G. K., Shen, X., Bir, S. C., & Kevil, C. G. (2013). Hydrogen sulfide chemical biology: Pathophysiological roles and detection. *Nitric Oxide*, *35*, 5–20.
57. Seth, P., Hsieh, P. N., Jamal, S., Wang, L., Gygi, S. P., Jain, M. K., Coller, J., & Stamler, J. S. (2019). Regulation of MicroRNA machinery and development by interspecies S-nitrosylation. *Cell*, *176*(5), P1014–P1025.E12. https://doi.org/10.1016/j.cell.2019.01.037
58. Sunico, C. R., Portillo, F., González-Forero, D., & Moreno-López, B. (2005). Nitric oxide–directed synaptic remodeling in the adult mammal CNS. *Journal of Neuroscience*, *25*(6), 1448–1458.
59. Bschor, T., & Bauer, M. (2006). Efficacy and mechanisms of action of lithium augmentation in refractory major depression. *Current Pharmaceutical Design*, *12*(23), 2985–2992.
60. Lu, Y.-R., Zhang, Y., Rao, Y.-B., Chen, X., Lou, H.-F., Zhang, Y., Xie, H.-Y., Fang, P., & Hu, L.-W. (2018). The changes in, and relationship between, plasma nitric oxide and corticotropin-releasing hormone in patients with

major depressive disorder. *Clinical and Experimental Pharmacology and Physiology*, *45*(1), 10–15.

61. Zhang, X. R., Wang, Y. X., Zhang, Z. J., Li, L., & Reynolds, G. P. (2012). The effect of chronic antipsychotic drug on hypothalamic expression of neural nitric oxide synthase and dopamine D2 receptor in the male rat. *PLOS ONE*, *7*(4), article e33247.
62. Ahmad, A., Dempsey, S. K., Daneva, Z., Azam, M., Li, N., Li, P.-L., & Ritter, J. K. (2018). Role of nitric oxide in the cardiovascular and renal systems. *International Journal of Molecular Sciences*, *19*(9), 2605. https://doi.org/10.3390/ijms19092605
63. Tribble, G. D., Angelov, N., Weltman, R., Wang, B.-Y., Eswaran, S. V., Gay, I. C., Parthasarathy, K., Dao, D.-H. V., Richardson, K. N., Ismail, N. M., Sharina, I. G., Hyde, E. R., Ajami, N. J., Petrosino, J. F., & Bryan, N. S. (2019). Frequency of tongue cleaning impacts the human tongue microbiome composition and enterosalivary circulation of nitrate. *Frontiers in Cellular and Infection Microbiology*, *9*, 39. https://doi.org/10.3389/fcimb.2019.00039
64. Fauste, E., Donis, C., Panadero, M. I., Otero, P., & Bocos, C. (2021, June 1). Fructose consumption hampers casotransmitter production. *Academia Letters*, article 1380. https://doi.org/10.20935/AL1380.3
65. Queiroga, C.S.F., Vercelli, A., & Vieira, H.L.A. (2015). Carbon monoxide and the CNS: challenges and achievements. *British Journal of Pharmacology*, *172*(6), 1533–1545.
66. Dreyer-Andersen, N., Almeida, A. S., Jensen, P., Kamand, M., Okarmus, J., Rosenberg, T., Friis, S. D., Martínez Serrano, A., Blaabjerg, M., Kristensen, B. W., Skrydstrup, T., Gramsbergen, J. B., Vieira, H.L.A., & Meyer, M. (2018). Intermittent, low dose carbon monoxide exposure enhances survival and dopaminergic differentiation of human neural stem cells. *PLOS ONE*, *13*(1), e0191207.
67. Trentini, J. F., O'Neill, J. T., Poluch, S., & Juliano, S. L. (2016). Prenatal carbon monoxide impairs migration of interneurons into the cerebral cortex. *Neurotoxicology*, *53*, 31–44.

CHAPTER 3: I HAVE A CRYSTAL BALL—AND IT'S YOUR GUT WALL

1. Wastyk, H. C., Fragiadakis, G. K., Perelman, D., Dahan, D., Merrill, B. D., Yu, F. B., Topf, M., Gonzalez, C. G., Van Treuren, W., Han, S., Robinson, J. L., Elias, J. E., Sonnenburg, E. D., Gardner, C. D., & Sonnenburg, J. L. (2021). Gut-microbiota-targeted diets modulate human immune status. *Cell*, *184*(16), 4137–4153.e14. https://doi.org/10.1016/j.cell.2021.06.019
2. Ashrafian, F., Raftar, S.K.A., Shahryari, A., Behrouzi, A., Yaghoubfar, R., Lari, A., Moradi, H. R., Khatami, S., Omrani, M. D., Vaziri, F., Masotti, A., & Siadat, S. D. (2021). Comparative effects of alive and pasteurized *Akkermansia muciniphila* on normal diet–fed mice. *Scientific Reports*, *11*, article 1789. https://doi.org/10.1038/s41598-021-95738-5
3. Thorburn, A. (2008). Aptosis and autophagy: Regulatory connections between two supposedly different processes. *Apoptosis*, *13*(1), 1–9. https://doi.org/10.1007/s10495-007-0154-9

4. Kaiko, G. E., Ryu, S. H., Koues, O. I., Collins, P. L., Solnica-Krezel, L., Pearce, E. J., Pearce, E. L., Oltz, E. M., & Stappenbeck, T. S. (2016). The colonic crypt protects stem cells from microbiota-derived metabolites. *Cell*, *165*(7), 1708–1720.
5. Saffarian, A., Mulet, C., Regnault, B., Amiot, A., Tran-Van-Nhieu, J., Ravel, J., Sobhani, I., Sansonetti, P. J., & Pédron, T. (2019). Crypt- and mucosa-associated core microbiotas in humans and their alteration in colon cancer patients. *mBio*, *10*(4), article 01315-19. https://doi.org/10.1128/mBio.01315
6. Chromek, M., Arvidsson, I., & Karpman, D. (2012). The antimicrobial peptide cathelicidin protects mice from *Escherichia coli* O157:H7-mediated disease. *PLOS ONE*, *7*(10), e46476.
7. da Costa, J. P., Cova, M., Ferreira, R., & Vitorino, R. (2015). Antimicrobial peptides: An alternative for innovative medicines? *Applied Microbiology and Biotechnology*, *99*(5), 2023–2040.
8. Brandl, K., Plitas, G., Schnabl, B., DeMatteo, R. P., & Pamer, E. G. (2007). MyD88-mediated signals induce the bactericidal lectin RegIII gamma and protect mice against intestinal *Listeria monocytogenes* infection. *The Journal of Experimental Medicine*, *204*(8), 1891–1900.
9. Yao, X., Zhang, C., Xing, Y., Xue, G., Zhang, Q., Pan, F., Wu, G., Hu, Y., Guo, Q., Lu, A., Zhang, X., Zhou, R., Tian, Z., Zeng, B., Wei, H., Strober, W., Zhao, L., & Meng, G. (2017). Remodelling of the gut microbiota by hyperactive NLRP3 induces regulatory T cells to maintain homeostasis. *Nature Communications*, *8*(1), article 1896.
10. Dignass, A. U. (2001). Mechanisms and modulation of intestinal epithelial repair. *Inflammatory Bowel Diseases*, *7*(1), 68–77.
11. Furusawa, Y., Obata, Y., Fukuda, S., Endo, T. A., Nakato, G., Takahashi, D., Nakanishi, Y., Uetake, C., Kato, K., Kato, T., Takahashi, M., Fukuda, N., Murakami, S., Miyauchi, E., Hino, S., Atarashi, K., Onawa, S., Fujimura, Y., Lockett, T., . . . Ohno, H. (2013). Commensal microbe-derived butyrate induces the differentiation of colonic regulatory T cells. *Nature*, *504*(7480), 446–450.
12. Zaborin, A., Krezalek, M., Hyoju, S., Defazio, J. R., Setia, N., Belogortseva, N., Bindokas, V. P., Guo, Q., Zaborina, O., & Alverdy, J. C. (2017). Critical role of microbiota within cecal crypts on the regenerative capacity of the intestinal epithelium following surgical stress. *AJP Gastrointestinal and Liver Physiology*, *312*(2), G112–G122.
13. Cantorna, M. T., Lin, Y.-D., Arora, J., Bora, S., Tian, Y., Nichols, R. G., & Patterson, A. D. (2019). Vitamin D regulates the microbiota to control the numbers of RORγt/FoxP3+ regulatory T cells in the colon. *Frontiers in Immunology*, *10*, 1772.
14. Kong, J., Zhang, Z., Musch, M. W., Ning, G., Sun, J., Hart, J., Bissonnette, M., & Li, Y. C. (2008). Novel role of the vitamin D receptor in maintaining the integrity of the intestinal mucosal barrier. *AJP Gastrointestinal and Liver Physiology*, *294*(1), G208–G216.
15. Cantarel, B. L., Waubant, E., Chehoud, C., Kuczynski, J., DeSantis, T. Z., Warrington, J., Venkatesan, A., Fraser, C. M., & Mowry, E. M. (2015). Gut

microbiota in multiple sclerosis: Possible influence of immunomodulators. *Journal of Investigative Medicine*, *63*(5), 729–734.

16. Zittermann, A., Ernst, J. B., Gummert, J. F., & Borgermann, J. (2014). Vitamin D supplementation, body weight and human serum 25-hydroxyvitamin D response: A systematic review. *European Journal of Nutrition*, *53*(2), 367–374.
17. Ghosh, S. S., Wang, J., Yannie, P. J., & Ghosh, S. (2020). Intestinal barrier dysfunction, LPS translocation, and disease development. *Journal of the Endocrine Society*, *4*(2), bvz039.
18. Liu, W., Hu, D., Huo, H., Zhang, W., Adiliaghdam, F., Morrison, S., Ramirez, J. M., Gul, S. S., Hamarneh, S. R., & Hodin, R. A. (2016). Intestinal alkaline phosphatase regulates tight junction protein levels. *Journal of the American College of Surgeons*, *222*(6), 1009–1017. https://doi.org/10.1016/j.jamcollsurg.2015.12.006
19. Malo, M. S., Alam, S. N., Mostafa, G., Zeller, S. J., Johnson, P. V., Mohammad, N., Chen, K. T., Moss, A. K., Ramasamy, S., Faruqui, A., Hodin, S., Malo, P. S., Ebrahimi, F., Biswas, B., Narisawa, S., Millán, J. L., Warren, H. S., Kaplan, J. B., Kitts, C. L., . . . Hodin, R. A. (2010). Intestinal alkaline phosphatase preserves the normal homeostasis of gut microbiota. *Gut*, *59*(11), 1476–1484. https://doi.org/10.1136/gut.2010.211706
20. Ghosh, S. S. Wang, J., Yannie, P. J., Cooper, R. C., Sandhu, Y. K., Kakiyama, G., Korzun, W. J., & Ghosh, S. (2021). Over-expression of intestinal alkaline phosphatase attenuates atherosclerosis. *Circulation Research*, *128*(11), 1646–1659. https://doi.org/10.1161/CIRCRESAHA.120.317144
21. Bates, J. M., Mittge, E., Kuhlman, J., Baden, K. N., Cheesman, S. E., & Guillemin, K. (2006). Distinct signals from the microbiota promote different aspects of zebrafish gut differentiation. *Developmental Biology*, *297*(2), 374–386. https://doi.org/10.1016/j.ydbio.2006.05.006
22. Kühn, F., Adiliaghdam, F., Cavallaro, P. M., Hamarneh, S. R., Tsurumi, A., Hoda, R. S., Munoz, A. R., Dhole, Y., Ramirez, J. M., Liu E., Vasan, R., Liu, Y., Samarbafzadeh, E., Nunez, R. A., Farber, M. Z., Chopra, V., Malo, M. S., Rahme, L. G., & Hodin, R. A. (2020). Intestinal alkaline phosphatase targets the gut barrier to prevent aging. *JCI Insight*, *5*(6), e134049. https://doi.org/10.1172/jci.insight.134049
23. Ghosh, S. S., Gehr, T. W., & Ghosh, S. (2014). Curcumin and chronic kidney disease (CKD): Major mode of action through stimulating endogenous intestinal alkaline phosphatase. *Molecules*, *19*(12), 20139–20156. https://doi.org/10.3390/molecules191220139
24. Ermolenko, E., Gromova, L., Borschev, Y., Voeikova, A., Karaseva, A., Ermolenko, K., Gruzdkov, A., & Suvorov, A. (2013). Influence of different probiotic lactic acid bacteria on microbiota and metabolism of rats with dysbiosis. *Bioscience of Microbiota, Food and Health*, *32*(2), 41–49. https://doi.org/10.12938/bmfh.32.41
25. Navis, M., Muncan, V., Sangild, P. T., Møller Willumsen, L., Koelink, P. J., Wildenberg, M. E., Abrahamse, E., Thymann, T., van Elburg, R. M., & Renes, I. B. (2020). Beneficial effect of mildly pasteurized whey protein on intestinal

integrity and innate defense in preterm and near-term piglets. *Nutrients*, *12*(4), 1125. https://doi.org/10.3390/nu12041125

26. Zhong, Z., Wheeler, M. D., Li, X., Froh, M., Schemmer, P., Yin, M., Bunzendaul, H., Bradford, B., & Lemasters, J. J. (2003). L-glycine: A novel anti-inflammatory, immunomodulatory, and cytoprotective agent. *Current Opinion in Clinical Nutrition & Metabolic Care*, *6*(2), 229–240. https://doi.org/10.1097/00075197-200303000-00013
27. Guzman-Stein, G., Bonsack, M., Liberty, J., & Delaney, J. P. (1989). Abdominal radiation causes bacterial translocation. *The Journal of Surgical Research*, *46*(2), 104–107.
28. Xavier, R. J., & Podolsky, D. K. (2007). Unravelling the pathogenesis of inflammatory bowel disease. *Nature*, *448*(7512), 427–434.
29. Diestel, C. F., Marques, R .G., Lopes-Paulo, F., Paiva, D., Horst, N. L., Caetano, C. E., & Portela, M. C. (2007). Role of L-glutamine and glycine supplementation on irradiated colonic wall. *International Journal of Colorectal Disease*, *22*(12), 1523–1529.
30. Cruz, M., Maldonado-Bernal, C., Mondragón-Gonzalez, R., Sanchez-Barrera, R., Wacher, N. H., Carvajal-Sandoval, G., & Kumate, J. (2008). Glycine treatment decreases proinflammatory cytokines and increases interferon-gamma in patients with type 2 diabetes. *Journal of Endocrinological Investigation*, *31*(8), 694–699. https://doi.org/10.1007/BF03346417
31. Nguyen, D., Samson, S. L., Reddy, V. T., Gonzalez, E. V., & Sekhar, R. V. (2013). Impaired mitochondrial fatty acid oxidation and insulin resistance in aging: novel protective role of glutathione. *Aging Cell*, *12*(3), 415–425.
32. Franceschi, C., & Campisi, J. (2014). Chronic inflammation (inflammageing) and its potential contribution to age-associated diseases. *The Journals of Gerontology: Series A*, *69*(Suppl. 1), S4–S9.
33. Martínez-Augustin, O., Rivero-Gutiérrez, B., Mascaraque, C., & Sánchez de Medina, F. (2014). Food derived bioactive peptides and intestinal barrier function. *International Journal of Molecular Sciences*, *15*(12), 22857–22873. https://doi.org/10.3390/ijms151222857
34. Bannai, M., & Kawai, N. (2012). New therapeutic strategy for amino acid medicine: Glycine improves the quality of sleep. *Journal of Pharmacological Sciences*, *118*(2), 145–148.
35. Nowotarski, S. L., Woster, P. M., & Casero, R. A. (2013). Polyamines and cancer: Implications for chemotherapy and chemoprevention. *Expert Reviews in Molecular Medicine*, *15*, e3. https://doi.org/10.1017/erm.2013.3
36. Larqué, E., Sabater-Molina, M., & Zamora, S. (2007). Biological significance of dietary polyamines. *Nutrition*, *23*(1), 87–95. https://doi.org/10.1016/j.nut.2006.09.006
37. Gallego, C., Kumar, H., García-Mantrana, I., du Toit, E., Suomela, J. P., Linderborg, K. M., Zhang, Y., Isolauri, E., Yang, B., Salminen, S., & Collado, M. C. (2017). Breast milk polyamines and microbiota interactions: impact of mode of delivery and geographical location. *Annals of Nutrition & Metabolism*, *70*(3), 184–190. https://doi.org/10.1159/000457134

38. Tofalo, R., Cocchi, S., & Suzzi, G. (2019). Polyamines and gut microbiota. *Frontiers in Nutrition*, *6*, 16. https://doi.org/10.3389/fnut.2019.00016
39. Kalač, P. (2014). Health effects and occurrence of dietary polyamines: A review for the period 2005–mid 2013. *Food Chemistry*. *161*, 27–39. https://doi.org/10.1016/j.foodchem.2014.03.102
40. Peulen, O., Gharbi, M., Powroznik, B., & Dandrifosse, G. (2004). Differential effect of dietary spermine on alkaline phosphatase activity in jejunum and ileum of unweaned rats. *Biochimie*, *86*(7), 487–493. https://doi.org/10.1016/j.biochi.2004.06.002
41. Dandrifosse, G., Peulen, O., El Khefif, N., Deloyer, P., Dandrifosse, A. C., & Grandfils, C. H. (2000). Are milk polyamines preventive agents against food allergy? *The Proceedings of the Nutrition Society*, *59*(1), 81–86. https://doi.org/10.1017/S0029665100000100
42. Ali, M. A., Poortvliet, E., Strömberg, R., & Yngve, A. (2011). Polyamines in foods: Development of a food database. *Food & Nutrition Research*, *55*, 5572. https://doi.org/10.3402/fnr.v55i0.5572

CHAPTER 4: A PERFECT STORM

1. Grand View Research. (2023, June). *Plastic market size, share & trends analysis report by product (PE, PP, PU, PVC, PET, Polystyrene, ABS, PBT, PPO, Epoxy Polymers, LCP, PC, Polyamide), by application, by end-use, by region, and segment forecasts, 2023–2030*. https://www.grandviewresearch.com/industry-analysis/global-plastics-market
2. Lear, G., Kingsbury, J. M., Franchini, S., Gambarini, V., Maday, S.D.M., Wallbank, J. A., Weaver, L., & Pantos, O. (2021). Plastics and the microbiome: Impacts and solutions. *Environmental Microbiome*, *16*(1), 2. https://doi.org/10.1186/s40793-020-00371-w
3. Dorsey, E. R., Sherer, T., Okun, M. S., & Bloem, B. R. (2018). The emerging evidence of the Parkinson pandemic. *Journal of Parkinson's Disease*, *8*(Suppl. 1), S3–S8. https://doi.org/10.3233/JPD-181474
4. Chua, K.-P., Fischer, M. A., & Linder, J. A. (2019). Appropriateness of outpatient antibiotic prescribing among privately insured US patients: ICD-10-CM based cross sectional study. *The British Medical Journal*, *364*, k592. https://doi.org/10.1136/bmj.k5092
5. Aminov, R. I. (2010). A brief history of the antibiotic era: Lessons learned and challenges for the future. *Frontiers in Microbiology*, *1*, 134. https://doi.org/10.3389/fmicb.2010.00134
6. World Health Organization. (2020, July 31). *Antibiotic resistance*. https://www.who.int/news-room/fact-sheets/detail/antibiotic-resistance
7. Garofalo, C., Vignaroli, C., Zandri, G., Aquilanti, L., Bordoni, D., Osimani, A., Clementi, F., & Biavasco, F. (2007). Direct detection of antibiotic resistance genes in specimens of chicken and pork meat. *International Journal of Food Microbiology*, *113*(1), 75–83.
8. Dubourg, G., Lagier, J. C., Robert, C., Armougom, F., Hugon, P., Metidji, S., Dione, N., Dangui, N. P., Pfleiderer, A., Abrahao, J., Musso, D., Papazian, L.,

Brouqui, P., Bibi, F., Yasir, M., Vialettes, B., & Raoult, D. (2014). Culturomics and pyrosequencing evidence of the reduction in gut microbiota diversity in patients with broad-spectrum antibiotics. *International Journal of Antimicrobial Agents*, *44*(2), 117–124. https://doi.org/10.1016/j.ijantimicag.2014.04.020

9. Theriot, C. M., Bowman, A. A., & Young, B. (2014). Antibiotic-induced shifts in the mouse gut microbiome and metabolome increase susceptibility to *Clostridium difficile* infection. *Nature Communications*, *5*, article 3114. https://doi.org/10.1128/mSphere.00045-15
10. Dethlefsen, L., & Relman, D. A. (2011). Incomplete recovery and individualized responses of the human distal gut microbiota to repeated antibiotic perturbation. *Proceedings of the National Academy of Sciences of the United States of America*, *108*(Suppl. 1), 4554–4561. https://doi.org/10.1073/pnas.1000087107
11. Anthony, W. E., Wang, B., Sukhum, K. V., D'Souza, A. W., Hink, T., Cass, C., Seiler, S., Reske, K. A., Coon, C., Dubberke, E. R., Burnham, C.-A. D., Dantas, G., & Kwon, J. H. (2022). Acute and persistent effects of commonly used antibiotics on the gut microbiome and resistome in healthy adults. *Cell Reports*, *39*(2), 110649. https://doi.org/10.1016/j.celrep.2022.110649
12. Haak, B. W., Lankelma, J. M., Hugenholtz, F., Belzer, C., de Vos, W. M., & Wiersinga, W. J. (2019). Long-term impact of oral vancomycin, ciprofloxacin and metronidazole on the gut microbiota in healthy humans, *Journal of Antimicrobial Chemotherapy*, *74*(3), 782–786.
13. Tapiainen, T., Koivusaari, P., Brinkac, L., Lorenzi, H. A., Salo, J., Renko, M., Pruikkonen, H., Pokka, T., Li, W., Nelson, K., Pirttilä, A. M., & Tejesvi, M. V. (2019). Impact of intrapartum and postnatal antibiotics on the gut microbiome and emergence of antimicrobial resistance in infants. *Scientific Reports*, *9*(1), 10635. https://doi.org/10.1038/s41598-019-46964-5
14. Candon, S., Perez-Arroyo, A., Marquet, C., Valette, F., Foray, A. P., Pelletier, B., Milani, C., Ventura, M., Bach, J. F., & Chatenoud, L. (2015). Antibiotics in early life alter the gut microbiome and increase disease incidence in a spontaneous mouse model of autoimmune insulin-dependent diabetes. *PLOS ONE*, *10*(5), 1–16. https://doi.org/10.1371/journal.pone.0125448
15. Zhao, Y., Wu, J., Li, J. V., Zhou, N. Y., Tang, H., & Wang, Y. (2013). Gut microbiota composition modifies fecal metabolic profiles in mice. *Journal of Proteome Research*, *12*(6), 2987–2999. https://doi.org/10.1021/pr400263n
16. Willing, B. P., Russell, S. L., & Finlay, B. B. (2011). Shifting the balance: Antibiotic effects on host-microbiota mutualism. *Nature Reviews Microbiology*, *9*(4), 233–243. https://doi.org/10.1038/nrmicro2536
17. Kalghatgi, S., Spina, C. S., Costello, J. C., Liesa, M., Morones-Ramirez, J. R., Slomovic, S., Molina, A., Shirihai, O. S., & Collins, J. J. (2013). Bactericidal antibiotics induce mitochondrial dysfunction and oxidative damage in mammalian cells. *Science Translational Medicine*, *5*(192), 192ra85. https://doi.org/10.1126/scitranslmed.3006055
18. Morgun, A., Dzutsev, A., Dong, X., Greer, R. L., Sexton, D. J., Ravel, J., Schuster, M., Hsiao, W., Matzinger, P., & Shulzhenko, N. (2015). Uncovering effects of antibiotics on the host and microbiota using transkingdom gene networks. *Gut*, *64*(11), 1732–1743. https://doi.org/10.1136/gutjnl-2014-308820

19. Azad, M. B., Konya, T., Persaud, R. R., Guttman, D. S., Chari, R. S., Field, C. J., Sears, M. R., Mandhane, P. J., Turvey, S. E., Subbarao, P., Becker, A. B., Scott, J. A., & Kozyrskyj, A. L., CHILD Study Investigators. (2016). Impact of maternal intrapartum antibiotics, method of birth and breastfeeding on gut microbiota during the first year of life: A prospective cohort study. *BJOG: An International Journal of Obstetrics and Gynaecology*, *123*(6), 983–993. https://doi.org/10.1111/1471-0528.13601
20. Nyangahu, D. D., Lennard, K. S., Brown, B. P., Darby, M. G., Wendoh, J. M., Havyarimana, E., Smith, P., Butcher, J., Stintzi, A., Mulder, N., Horsnell, W., & Jaspan, H. B. (2018). Disruption of maternal gut microbiota during gestation alters offspring microbiota and immunity. *Microbiome*, *6*(1), 1–10. https://doi.org/10.1186/s40168-018-0511-7
21. Tapiainen, T., Koivusaari, P., Brinkac, L., Lorenzi, H. A., Salo, J., Renko, M., Pruikkonen, H., Pokka, T., Li, W., Nelson, K., Pirttilä, A. M., & Tejesvi, M. V. (2019). Impact of intrapartum and postnatal antibiotics on the gut microbiome and emergence of antimicrobial resistance in infants. *Scientific Reports*, *9*(1), article 10635. https://doi.org/10.1038/s41598-019-46964-5
22. Dobbler, P., Mai, V., Procianoy, R. S., Silveira, R. C., Corso, A. L., & Roesch, L. (2019). The vaginal microbial communities of healthy expectant Brazilian mothers and its correlation with the newborn's gut colonization. *World Journal of Microbiology and Biotechnology*, *35*(10), 1–14. https://doi.org/10.1007/s11274-019-2737-3
23. Czeizel, A. E., Rockenbauer, M., Sørensen, H. T., & Olsen, J. (2001). The teratogenic risk of trimethoprim-sulfonamides: A population based case-control study. *Reproductive Toxicology*, *15*, 637–646. https://doi.org/10.1016/S0890-6238(01)00178-2
24. Stokholm, J., Sevelsted, A., Bønnelykke, K., & Bisgaard, H. (2014). Maternal propensity for infections and risk of childhood asthma: A registry-based cohort study. *The Lancet Respiratory Medicine*, *2*(8), 631–637. https://doi.org/10.1016/S2213-2600(14)70152-3
25. Kenyon, S., Pike, K., Jones, D. R., Brocklehurst, P., Marlow, N., Salt, A., & Taylor, D. J. (2008). Childhood outcomes after prescription of antibiotics to pregnant women with spontaneous preterm labour: 7-year follow-up of the ORACLE II trial. *The Lancet*, *372*(9646), 1319–1327. https://doi.org/10.1016/S0140-6736(08)61203-9
26. Mueller, N. T., Whyatt, R., Hoepner, L., Oberfield, S., Dominguez-Bello, M. G., Widen, E. M., Hassoun, A., Perera, F., & Rundle, A. (2015). Prenatal exposure to antibiotics, cesarean section and risk of childhood obesity. *International Journal of Obesity*, *39*(4), 665–670. https://doi.org/10.1038/ijo.2014.180
27. Tormo-Badia, N., Håkansson, Å., Vasudevan, K., Molin, G., Ahrné, S., & Cilio, C. M. (2014). Antibiotic treatment of pregnant non-obese diabetic mice leads to altered gut microbiota and intestinal immunological changes in the offspring. *Scandinavian Journal of Immunology*, *80*(4), 250–260. https://doi.org/10.1111/sji.12205
28. Källén, B., & Danielsson, B. R. (2014). Fetal safety of erythromycin. An update

of Swedish data. *European Journal of Clinical Pharmacology*, *70*(3), 355–360. https://doi.org/10.1007/s00228-013-1624-3

29. Crider, K. S., Cleves, M. A., Reefhuis, J., Berry, R. J., Hobbs, C. A., & Hu, D. J. (2009). Antibacterial medication use during pregnancy and risk of birth defects. *Archives of Pediatrics and Adolescent Medicine*, *163*(11), 978–985. https://doi.org/10.1001/archpediatrics.2009.188
30. Hermansson, H., Kumar, H., Collado, M. C., Salminen, S., Isolauri, E., & Rautava, S. (2019). Breast milk microbiota is shaped by mode of delivery and intrapartum antibiotic exposure. *Frontiers in Nutrition*, *6*, 475. https://doi.org/10.3389/fnut.2019.00004
31. Korpela, K., Salonen, A., Virta, L. J., Kekkonen, R. A., Forslund, K., Bork, P., & de Vos, W. M. (2016). Intestinal microbiome is related to lifetime antibiotic use in Finnish pre-school children. *Nature Communications*, *7*, 1–8. https://doi.org/10.1038/ncomms10410
32. Ni, J., Friedman, H., Boyd, B. C., McGurn, A., Babinski, P., Markossian, T., & Dugas, L. R. (2019). Early antibiotic exposure and development of asthma and allergic rhinitis in childhood. *BMC Pediatrics*, *19*(1), 1–8. https://doi.org/10.1186/s12887-019-1594-4
33. Yamamoto-Hanada, K., Yang, L., Narita, M., Saito, H., & Ohya, Y. (2017). Influence of antibiotic use in early childhood on asthma and allergic diseases at age 5. *Annals of Allergy, Asthma & Immunology*, *119*(1), 54–58. https://doi.org/10.1016/j.anai.2017.05.013
34. Bailey, L. C., Forrest, C. B., Zhang, P., Richards, T. M., Livshits, A., & DeRusso, P. A. (2014). Association of antibiotics in infancy with early childhood obesity. *JAMA Pediatrics*, *168*(11), 1063–1069. https://doi.org/10.1001/jamapediatrics.2014.1539
35. Grube, A., Donaldson, D., Kiely, T., & Wu, L. (2011). Pesticides industry sales and usage: 2006 and 2007 market estimates. https://www.epa.gov/sites/default/files/2015-10/documents/market_estimates2007.pdf
36. Canadian Food Inspection Agency. (2017). Safeguarding with science: Glyphosate testing in 2015–2016. https://inspection.canada.ca/food-safety-for-industry/food-chemistry-and-microbiology/food-safety-testing-bulletin-and-reports/executive-summary/glyphosate-testing/eng/1491846907641/1491846907985
37. IARC Working Group on the Evaluation of Carcinogenic Risks to Humans. (2015). *IARC monographs, volume 112: Some organophosphate insecticides and herbicides*. World Health Organization. https://monographs.iarc.who.int/wp-content/uploads/2018/07/mono112.pdf
38. Mesnage, R., Calatayud, M., Duysburgh, C., Marzorati, M., & Antoniou, M. N. Alterations in human gut microbiome composition and metabolism after exposure to glyphosate and Roundup and/or a spore-based formulation using the SHIME technology. *Gut Microbiome*, *3*, e6.
39. El-Shenawy, N. S. (2009). Oxidative stress responses of rats exposed to Roundup and its active ingredient glyphosate. *Environmental Toxicology and Pharmacology*, *28*(3), 379–385. https://doi.org/10.1016/j.etap.2009.06.001

40. Lushchak, O. V., Kubrak, O. I., Storey, J. M., Storey, K. B., & Lushchak, V. I. (2009). Low toxic herbicide Roundup induces mild oxidative stress in goldfish tissues. *Chemosphere*, *76*(7), 932–937. https://doi.org/10.1016/j.chemosphere.2009.04.045
41. Strandwitz, P., Kim, K. H., Terekhova, D., Liu, J. K., Sharma, A., Levering J., McDonald, D., Dietrich, D., Ramadhar, T. R., Lekbua, A., Mroue, N., Liston, C., Stewart, E. J., Dubin, M. J., Zengler, K., Knight, R., Gilbert, J. A., Clardy, J., & Lewis, K. (2019). GABA-modulating bacteria of the human gut microbiota. *Nature Microbiology*, *4*(3), 396–403. https://doi.org/10.1038/s41564-018-0307-3
42. Briguglio, M., Dell'Osso, B., Panzica, G., Malgaroli, A., Banfi, G., Dina, C. Z., Galentino, R., & Porta, M. (2018). Dietary neurotransmitters: A narrative review on current knowledge. *Nutrients*, *10*(5), 591. https://doi.org/10.3390/nu10050591
43. Zhang, L. S., & Davies, S. S. (2016). Microbial metabolism of dietary components to bioactive metabolites: Opportunities for new therapeutic interventions. *Genome Medicine*, *8*, 46. https://doi.org/10.1186/s13073-016-0296-x
44. Hill-Burns, E. M., Debelius, J. W., Morton, J. T., Wissemann, W. T., Lewis, M. R., Wallen, Z. D., Peddada, S. D., Factor, S. A., Molho, E., Zabetian, C. P., Knight, R., & Payami, H. (2017). Parkinson's disease and Parkinson's disease medications have distinct signatures of the gut microbiome. *Movement Disorders*, *32*(5), 739–749. https://doi.org/10.1002/mds.26942
45. Nguyen, T. T., Hathaway, H., Kosciolek, T., Knight, R., & Jeste, D. V. (2019). Gut microbiome in serious mental illnesses: A systematic review and critical evaluation. *Schizophrenia Research*, *234*, 24–40. https://doi.org/10.1016/j.schres.2019.08.026
46. Chen, J.-J., Zheng, P., Liu, Y.-Y., Zhong, X.-G., Wang, H.-Y., Guo, Y.-J., Xie, P. (2018). Sex differences in gut microbiota in patients with major depressive disorder. *Neuropsychiatric Disease and Treatment*, *14*, 647–655. https://doi.org/10.2147/NDT.S159322
47. La Merrill, M. A., Vandenburg, L. N., Smith, M. T., Goodson, W., Browne, P., Patisaul, H. B., Guyton, K. Z., Kortenkamp, A., Cogliano, V. J., Woofruff, T. J., Rieswijk, L., Sone, H., Korach, K. S., Gore, A. C., Zeise, L., & Zoeller, R. T. (2020). Consensus on the key characteristics of endocrine-disrupting chemicals as a basis for hazard identification. *Nature Reviews Endocrinology*, *16*(1), 45–57. https://doi.org/10.1038/s41574-019-0273-8
48. Brucker-Davis, F., Thayer, K., & Colborn, T. (2001). Significant effects of mild endogenous hormonal changes in humans: Considerations for low-dose testing. *Environmental Health Perspectives*, *109*(Suppl. 1), 21–26. https://doi.org/10.1289/ehp.01109s121
49. Hampl, R., & Stárka, L. (2020). Endocrine disruptors and gut microbiome interactions. *Physiological Research*, *69*(Suppl. 2), S211–S223. https://doi.org/10.33549/physiolres.934513
50. Winkler, J., Liu, P., Phong, K., Hinrichs, J. H., Ataii, N., Williams, K., Hadler-Olsen, E., Samson, S., Gartner, Z. J., Fisher, S., & Werb, Z. (2022). Bisphenol A replacement chemicals, BPF and BPS, induce protumorigenic changes in human

mammary gland organoid morphology and proteome. *Proceedings of the National Academy of Sciences of the United States of America*, *119*(11), e2115308119.

51. Evariste, L., Barret, M., Mottier, A., Mouchet, F., Gauthier, L., & Pinelli, E. (2019). Gut microbiota of aquatic organisms: A key endpoint for ecotoxicological studies. *Environmental Pollution*, *248*, 989–999. https://doi.org/10.1016/j.envpol.2019.02.101
52. Nowak, K., Jabłońska, E., & Rataczak-Wrona, W. (2019). Immunomodulatory effects of synthetic endocrine disrupting chemicals on the development and functions of human immune cells. *Environment International*, *125*, 350–364. https://doi.org/10.1016/j.envint.2019.01.078
53. Hampl, R., & Starka, L. (2020). Endocrine disruptors and gut microbiome interactions. *Physiological Research*, *69*(Suppl. 2), S211–S223. https://doi.org/10.33549/physiolres.934513
54. De Punder, K., & Pruimboom, L. (2013). The dietary intake of wheat and other cereal grains and their role in inflammaton. *Nutrients*, *5*(3), 771–787. https://doi.org/10.3390/nu5030771
55. Schumacher, U., Gräfin von Armansperg, N., Kreipe, H., & Welsch, U. (1996). Lectin binding and uptake in human (myelo)monocytic cell lines: HL60 and U937. *Ultrastructural Pathology*, *20*(5), 463–471. https://doi.org/10.3109/01913129609016350
56. Kataoka, H., Ushiyama, A., Kawakami, H., Akimoto, Y., Matsubara, S., & Iijim, T. (2016). Fluorescent imaging of endothelial glycocalyx layer with wheat germ agglutinin using intravital microscopy. *Microscopy Research and Technique*, *79*(1), 31–37. https://doi.org/10.1002/jemt.22602
57. Mochizuki, H., Fukui, M., Hatou, S., Yamada, M., & Tsubota, K. (2010). *Clinical Ophthalmology*, *4*, 925–930. https://doi.org/10.2147/opth.s12648
58. Plattner, V. E., Germann, B., Neuhaus, W., Noe, C. R., Gabor, F., & Wirth, M. (2010). Characterization of two blood-brain barrier mimicking cell lines: Distribution of lectin-binding sites and perspectives for drug delivery. *International Journal of Pharmaceutics*, *387*(1–2), 34–41. https://doi.org/10.1016/j.ijpharm.2009.11.030
59. Somasundaram, S., Rafi, S., Hayllar, J., Sigthorsson, G., Jacob, M., Price, A., Macpherson, A., Mahmod, T., Scott, D., Wrigglesworth, J., & Bjarnason, I. (1997). Mitochondrial damage: A possible mechanism of the "topical" phase of NSAID induced injury to the rat intestine. *Gut*, *41*(3), 344–353.
60. Rogers, M.A.M., & Aronoff, D. M. (2016). The influence of non-steroidal anti-inflammatory drugs on the gut microbiome. *Clinical Microbiology and Infection*, *22*(2), 178.E1–178.E9. https://doi.org/10.1016/j.cmi.2015.10.003
61. Somasundaram, S., Rafi, S., Hayllar, J., Sigthorsson, G., Jacob, M., Price, A., Macpherson, A., Mahmod, T., Scott, D., Wrigglesworth, J., & Bjarnason, I. (1997). Mitochondrial damage: A possible mechanism of the "topical" phase of NSAID induced injury to the rat intestine. *Gut*, *41*(3), 344–353.
62. Coxib and Traditional NSAID Trialists' (CNT) Collaboration. (2013). Vascular and upper gastrointestinal effects of non-steroidal anti-inflammatory drugs: Meta-analyses of individual participant data from randomised trials. *The Lancet*, *382*(9894), 769–779.

63. Neuroscience News. (2022, November 21). NSAIDs may worsen arthritis inflammation. https://neurosciencenews.com/nsaids-arthritis-inflammation-21905/
64. Riordan, S. M., McIver, C. J., Wakefield, D., Thomas, M. C., Duncombe, V. M., & Bolin, T. D. (1999). Serum immunoglobulin and soluble IL-2 receptor levels in small intestinal overgrowth with indigenous gut flora. *Digestive Diseases and Sciences*, *44*(5), 939–944.
65. Jackson, M. A., Goodrich, J. K., Maxan, M. E., Freedberg, D. E., Abrams, J. A., Poole, A. C., Sutter, J. L., Welter, D., Ley, R. E., Bell, J. T., Spector, T. D., & Steves, C. J. (2016). Proton pump inhibitors alter the composition of the gut microbiota. *Gut*, *65*(5), 749–756. https://doi.org/10.1136/gutjnl-2015-310861
66. Clooney, A. G., Bernstein, C. N., Leslie, W. D., Vagianos, K., Sargent, M., Laserna-Mendieta, E. J., Claesson, M. J., & Targownik, L. E. (2016). A comparison of the gut microbiome between long-term users and non-users of proton pump inhibitors. *Alimentary Pharmacology & Therapeutics*, *43*(9), 974–984.
67. Haenisch, B., von Holt, K., Wiese, B., Prokein, J., Lange, C., Ernst, A., Brettschneider, C., König, H.-H., Werle, J., Weyerer, S., Luppa, M., Riedel-Heller, S. G., Fuchs, A., Pentzek, M., Weeg, D., Bickel, H., Broich, K., Jessen, F., Wolfgang, M., & Scherer, M. (2015). Risk of dementia in elderly patients with the use of proton pump inhibitors. *European Archives of Psychiatry and Clinical Neuroscience*, *265*(5), 419–428.
68. Bateman, B. T., Bykov, K., Choudhry, N. K., Schneeweiss, S., Gagne, J. J., Polinski, J. M., Franklin, J. M., Doherty, M., Fischer, M. A., & Rassen, J. A. (2013). Type of stress ulcer prophylaxis and risk of nosocomial pneumonia in cardiac surgical patients: Cohort study. *The British Medical Journal*, *347*, f5416.
69. Klatte, D.C.F., Gasparini, A., Xu, H., de Deco, P., Trevisan, M., Johansson, A.L.V., Wettermark, B., Ärnlöv, J., Janmaat, C. J., Lindholm, B., Dekker, F. W., Coresh, J., Grams, M. E., & Carrero, J. J. (2017). Association between proton pump inhibitor use and risk of progression of chronic kidney disease. *Gastroenterology*, *153*(3), 702–710.
70. Shah, N. H., LePendu, P., Bauer-Mehren, A., Ghebremariam, Y. T., Iyer, S. V., Marcus, J., Nead, K. T., Cooke, J. P., & Leeper, N. J. (2015). Proton pump inhibitor usage and the risk of myocardial infarction in the general population. *PLOS ONE*, *10*(6), e0124653.
71. Freedberg, D. E., Kim, L. S., & Yang, Y.-X. (2017). The risks and benefits of long-term use of proton pump inhibitors: Expert review and best practice advice from the American Gastroenterological Association. *Gastroenterology*, *152*(4), 706–715.
72. Cox, K. D., Covernton, G. A., Davies, H. L., Dower, J. F., Juanes, F., & Dudas, S. E. (2019). Human consumption of microplastics. *Environmental Science & Technology*, *53*(12), 7068–7074.
73. Buhyan, S. (2022). Effects of microplastics on fish and in human health. *Frontiers in Environmental Science*, *10*, article 827289. https://doi.org/10.3389/fenvs.2022.827289
74. Schwabl, P., Köppel, S., Königshofer, P., Bucsics, T., Trauner, M., Reiberger, T., & Liebmann, B. (2019). Detection of various microplastics in human stool: A

Prospective Case Series. *Annals of Internal Medicine*, *171*(7), 453–457. https://doi.org/10.7326/M19-0618

75. Tamargo, A., Molinero, N., Reinosa, J. J., Alcolea-Rodriguez, V., Portela, R., Banares, M. A., Fernandez, J. F., & Moreno-Arribas, M. V. (2022). PET microplastics affect human gut microbiota communities during simulated gastrointestinal digestion, first evidence of plausible polymer biodegradation during human digestion. *Scientific Reports*, *12*, 528. https://doi.org/10.1038/s41598-021-04489-w
76. Tamargo, A., Molinero, N., Reinosa, J. J., Alcolea-Rodriguez, V., Portela, R., Bañares, M. A., Fernández, J. F., & Moreno-Arribas, M. V. (2022). PET microplastics affect human gut microbiota communities during simulated gastrointestinal digestion, first evidence of plausible polymer biodegradation during human digestion. *Scientific Reports*, *12*, 528. https://doi.org/10.1038/s41598-021-04489-w
77. Zhang, X., Wang, H., Peng, S., Kang, J., Xie, Z., Tang, R., Xing, Y., He, Y., Yuan, H., Xie, C., & Liu, Y. (2022). Effect of microplastics on nasal and intestinal microbiota of the high-exposure population. *Frontiers in Public Health*, *10*, article 1005535. https://doi.org/10.3389/fpubh.2022.1005535
78. Deng, Y., Zhang, Y., Lemos, B., & Ren, H. (2017). Tissue accumulation of microplastics in mice and biomarker responses suggest widespread health risks of exposure. *Scientific Reports*, *7*, article 46687. https://doi.org/10.1038/srep46687
79. Medley, E. A., Spratlen, M. J., Yan, B., Herbstman, J. B., & Deyssenroth, M. A. (2023, February 27). A systematic review of the placental translocation of micro- and nanoplastics. *Current Environmental Health Reports*. (Epub ahead of print.) https://doi.org/10.1007/s40572-023-00391-x

CHAPTER 5: HIPPOCRATES WAS RIGHT

1. McGee, M. D., Weber, D., Day, N., Vitelli, C., Crippen, D., Herndon, L. A., Hall, D. H., & Melov, S. (2011). Loss of intestinal nuclei and intestinal integrity in aging *C. elegans*. *Aging Cell*, *10*(4), 699–710. https://doi.org/10.1111/j.1474-9726.2011.00713.x
2. Gundry, S. R. (2018). Remission/cure of autoimmune diseases by a lectin limited diet supplemented with probiotics, prebiotics, and polyphenols, *Circulation*, *137*(Suppl. 1), abstract AP238.
3. Wang, X., Chen, Z., Qiao, S., Zhu, Q., Zuo, Z., & Guo, B. (2022). Analysis of alterations of the gut microbiota in moderate to severe psoriasis patients using 16S rRNA gene sequencing. *Indian Journal of Dermatology*, *67*(5), 495–503. https://doi.org/10.4103/ijd.ijd_297_22
4. Brewer, R. C., Lanz, T. V., Hale, C. R., Sepich-Poore, G. D., Martino, C., Swafford, A. D., Carroll, T. S., Kongpachith, S., Blum, L. K., Elliott, S. E., Blachere, N. E., Parveen, S., Fak, J., Yao, V., Troyanskaya, O., Frank, M. O., Bloom, M. S., Jahanbani, S., Gomez, A. M., . . . Orange, D. E. (2023). Oral mucosal breaks trigger anti-citrullinated bacterial and human protein antibody responses in rheumatoid arthritis. *Science Translational Medicine*, *15*(684), eabq8476. https://doi.org/10.1126/scitranslmed.abq8476

5. Yoon, H., Shaw, J. L., Haigis, M. C., & Greka, A. (2021). Lipid metabolism in sickness and in health: Emerging regulators of lipotoxicity. *Molecular Cell*, *81*(18), 3708–3730.
6. van den Munckhof, I.C.L., Kurilshikov, A., ter Horst, R., Riksen, N. P., Joosten, L.A.B., Zhernikova, J., Fu, J., Keating, S. T., Netea, M. G., de Graaf, J., & Rutten, J.H.W. (2018). Role of gut microbiota in chronic low-grade inflammation as potential driver for atherosclerotic cardiovascular disease: A systematic review of human studies. *Obesity Reviews*, *19*(12),1719–1734. https://doi.org/10.1111/obr.12750
7. Banks, W. A., Sharma, P., Bullock, K. M., Hansen, K. M., Ludwig, N., & Whiteside, T. L. (2020). Transport of extracellular vesicles across the blood-brain barrier: Brain pharmacokinetics and effects of inflammation. *International Journal of Molecular Sciences*, *21*(12), 4407. https://doi.org/10.3390/ijms21124407
8. Farnum, C. E., & Wilsman, N. J. (1984). Lectin-binding histochemistry of non-decalcified growth plate cartilage: A postembedment method for light microscopy of epon-embedded tissue. *The Journal of Histochemistry and Citochemistry*, *32*(6), 593–607.
9. Edfeldt, K., Swedenborg, J., Hansson, G. K., & Yan, Z.-Q. (2002). Expression of toll-like receptors in human atherosclerotic lesions: a possible pathway for plaque activation. *Circulation*, *105*(10), 1158–1161.
10. Carnevale, R., Nocella, C., Petrozza, V., Cammisotto, V., Pacini, L., Sorrentino, V., Martinelli, O., Irace, L., Sciarretta, S., Frati, G., Pastori, D., & Violi, F. (2018). Localization of lipopolysaccharide from *Escherichia coli* into human atherosclerotic plaque. *Scientific Reports*, *8*(1), 3598. https://doi.org/10.1038/s41598-018-22076-4
11. Aguilar, E. C., Santos, L. C., Leonel, A. J., de Oliveira, J. S., Santos, E. A., Navia-Pelaez, J. M., da Silva, J. F., Mendes, B. P., Capettini, L.S.A., Teixeira, L. G., Lemos, V. S., & Alvarez-Leite, J. I. (2016). Oral butyrate reduces oxidative stress in atherosclerotic lesion sites by a mechanism involving NADPH oxidase down-regulation in endothelial cells. *The Journal of Nutritional Biochemistry*, *34*, 99–105. https://doi.org/10.1016/j.jnutbio.2016.05.002
12. Ghosh, S. S., Wang, J., Yannie, P. J., Cooper, R. C., Sandhu, Y. K., Kakiyama, G., Korzun, W. J., & Ghosh, S. (2021). Over-expression of intestinal alkaline phosphatase attenuates atherosclerosis. *Circulation Research*, *128*(11), 1646–1659.
13. Lehtiniemi, J., Karhunen, P. J., Goebeler, S., Nikkari, S., & Nikkari, S. T. (2005). Identification of different bacterial DNAs in human coronary arteries. *European Journal of Clinical Investigation*, *35*(1), 13–16. https://doi.org/10.1111/j.1365-2362.2005.01440.x
14. Ott, S. J., El Mokhtari, N. E., Musfeldt, M., Hellmig, S., Freitag, S., Rehman, A., Kühbacher, T., Nikolaus, S., Namsolleck, P., Blaut, M., Hampe, J., Sahly, H., Reinecke, A., Haake, N., Günther, R., Krüger, D., Lins, M., Herrmann, G., Fölsch, U. R., . . . Schreiber, S. (2006). Detection of diverse bacterial signatures in atherosclerotic lesions of patients with coronary heart disease. *Circulation*, *113*(7), 929–937. https://doi.org/10.1161/CIRCULATIONAHA.105.579979

15. Qi, Y., Wu, H.-M., Yang, Z., Zhou, Y.-F., Jin, L., Yang, M.-F., & Wang, F.-Y. (2022). New insights into the role of oral microbiota dysbiosis in the pathogenesis of inflammatory bowel disease. *Digestive Diseases and Sciences*, *67*(1), 42–55. https://doi.org/10.1007/s10620-021-06837-2
16. Seymour, G. J., Ford, P. J., Cullinan, M. P., Leishman, S., & Yamazaki, K. (2007). Relationship between periodontal infections and systemic disease. *Clinical Microbiology and Infection*, *13*(Suppl. 4), 3–10.
17. Ridker, P. M, Bhatt, D., Pradhan, A., Glynn, R. J., MacFadyen, J. G., & Nissen, S. E. (2023). Inflammation and cholesterol as predictors of cardiovascular events among patients receiving statin therapy: A collaborative analysis of three randomised trials. *The Lancet*, *401*(10384), P1293–P1301. https://doi.org/10.1016/S0140-6736(23)00215-5
18. Gundry, S. R. (2019). Dietary lectins cause coronary artery disease via an autoimmune endothelial attack mediated by interleukin 16. *Arteriosclerosis, Thrombosis, and Vascular Biology*, *38*(Suppl. 1), abstract 412. https://doi.org/10.1161/atvb.38.suppl_1.412
19. Gundry, S. R. (2015). Twelve year followup for managing coronary artery disease using a nutrigenomics based diet and supplement program with quarterly assessment of biomarkers. *Arteriosclerosis, Thrombosis, and Vascular Biology*, *35*(Suppl. 1), abstract 309. https://doi.org/10.1161/atvb.35.suppl_1.309
20. Kawashima, H., Serruys, P. W., Ono, M., Hara, H., O'Leary, N., Mack, M. J., Holmes, D. R., Morice, M. C., Head, S. J., Kappetein, A. P., Thuijs, D.J.F.M., Milojevic, M., Noack, T., Mohr, F. W., Davierwala, P. M., Sharif, F., McEvoy, J. W., & Onuma, Y. (2021). Impact of optimal medical therapy on 10-year mortality after coronary revascularization. *Journal of the American College of Cardiology*, *78*(1), 27–38. https://doi.org/10.1016/j.jacc.2021.04.087
21. Methe, H., Kim, J. O., Kofler, S., Nabauer, M., & Weis, M. (2005). Statins decrease toll-like receptor 4 expression and downstream signaling in human CD14+ monocytes. *Arteriosclerosis, Thrombosis, and Vascular Biology*, *25*(7), 1439–1445. https://doi.org/10.1161/01.ATV.0000168410.44722.86
22. Zhang, Y., Zhang, S., Li, B., Luo, Y., Gong, Y., Jin, X., Zhang, J., Zhou, Y., Zhuo, X., Wang, Z., Zhao, X., Han, X., Gao, Y., Yu, H., Liang, D., Zhao, S., Sun, D., Wang, D., Xu, W., . . . Li, Y. (2022). Gut microbiota dysbiosis promotes age-related atrial fibrillation by lipopolysaccharide and glucose-induced activation of NLRP3-inflammasome. *Cardiovascular Research*, *118*(3) 785–797. https://doi.org/10.1093/cvr/cvab114
23. Li, J., Zhao, F., Wang, Y., Chen, J., Tao, J., Tian, G., Wu, S., Liu, W., Cui, Q., Geng, B., Zhang, W., Weldon, R., Auguste, K., Yang, L., Liu, X., Chen, L., Yang, X., Zhu, B., & Cai, J. (2017). Gut microbiota dysbiosis contributes to the development of hypertension. *Microbiome*, *5*(1), 14. https://doi.org/10.1186/s40168-016-0222-x
24. Li, Q., Gao, B., Siqin, B., He, Q., Zhang, R., Meng, X., Zhang, N., Zhang, N., & Li, M. (2021). Gut microbiota: A novel regulator of cardiovascular disease and key factor in the therapeutic effects of flavonoids. *Frontiers in Pharmacology*, *12*, article 651926. https://doi.org/10.3389/fphar.2021.651926
25. Li, J., Zhao, F., Wang, Y., Chen, J., Tao, J., Tian, G., Wu, S., Liu, W., Cui, Q.,

Geng, B., Zhang, W., Weldon, R., Auguste, K., Yang, L., Liu, X., Chen, L., Yang, X., Zhu, B., & Cai, J. (2017). Gut microbiota dysbiosis contributes to the development of hypertension. *Microbiome*, *5*(1), 14. https://doi.org/10.1186/s40168-016-0222-x

26. Brandsma, E., Kloosterhuis, N. J., Koster, M., Dekker, D. C., Gijbels, M.J.J., van der Velden, S., Ríos-Morales, M., van Fassen, M.J.R., Loreti, M. G., de Bruin, A., Fu, J., Kuipers, F., Bakker, B. M., Westerterp, M., de Winther, M.P.J., Hofker, M. H., van de Sluis, B., & Koonen, D.P.Y. (2019). A proinflammatory gut microbiota increases systemic inflammation and accelerates atherosclerosis. *Circulation Research*, *124*(1), 94–100.
27. Cani, P. D., Amar, J., Iglesias, M. A., Poggi, M., Knauf, C., Bastelica, D., Neyrinck, A. M., Fava, F., Tuohy, K. M., Chabo, C., Waget, A., Delmée, E., Cousin, B., Sulpice, T., Chamontin, B., Ferrières, J., Tanti, J.-F., Gibson, G. R., Casteilla, L., . . . Burcelin, R. (2007). Metabolic endotoxemia initiates obesity and insulin resistance. *Diabetes*, *56*(7), 1761–1772.
28. Trøseid, M., Nestvold, T. K., Rudi, K., Thoresen, H., Nielsen, E. W., & Lappegård, K. T. (2013). Plasma lipopolysaccharide is closely associated with glycemic control and abdominal obesity: Evidence from bariatric surgery. *Diabetes Care*, *36*(11), 3627–3632.
29. Sonnenburg, J. L., & Bäckhed, F. (2016). Diet-microbiota interactions as moderators of human metabolism. *Nature*, *535*(7610), 56–64.
30. Huang, Z. Y., Stabler, T., Pei, F. X., & Kraus, V. B. (2016). Both systemic and local lipopolysaccharide (LPS) burden are associated with knee OA severity and inflammation. *Osteoarthritis and Cartilege*, *24*(10), 1769–1775.
31. Ramasamy, B., Magne, F., Tripathy, S. K., Venugopal, G., Mukherjee, D., & Balamurugan, R. (2021). Association of gut microbiome and vitamin D deficiency in knee osteoarthritis patients: A pilot study. *Nutrients*, *13*(4), 1272. https://doi.org/10.3390/nu13041272
32. Li, R., Boer, C. G., Oei, L., & Medina-Gomez, C. (2021). The gut microbiome: A new frontier in musculoskeletal research. *Current Osteoporosis Reports*, *19*(3), 347–357. https://doi.org/10.1007/s11914-021-00675-x
33. Lei, M., Guo, C., Wang, D., Zhang, C., & Hua, L. (2017). The effect of probiotic *Lactobacillus casei* Shirota on knee osteoarthritis: A randomised double-blind, placebo-controlled clinical trial. *Beneficial Microbes*, *8*(5), 697–703.
34. Schott, E. M., Farnsworth, C. W., Grier, A., Lillis, J. A., Soniwala, S., Dadourian, G. H., Bell, R. D., Doolittle, M. L., Villani, D. A., Awad, H., Ketz, J. P., Kamal, F., Ackeret-Bicknell, C., Ashton, J. M., Gill, S. R., Mooney, R. A., & Zuscik, M. J. (2018). Targeting the gut microbiome to treat the osteoarthritis of obesity. *JCI Insight*, *3*(8), e95997.
35. Kare, S. K., Vinay, V., Maresz, K., Prisk, V., & Vik, H. (2022). *Tamarindus indica* seed extract–based botanical compositions alleviate knee pain and improve joint function in mild-to-moderate osteoarthritis: A randomized, double-blind, placebo-controlled clinical study. *Evidence-Based Complementary and Alternative Medicine*, *2022*, article 2226139. https://doi.org/10.1155/2022/2226139
36. Kasai, C., Sugimoto, K., Moritani, I., Tanaka, J., Oya, Y., Inoue, H., Tameda, M., Shiraki, K., Ito, M., Takei, Y., & Takase, K. (2015). Comparison of the gut

microbiota composition between obese and non-obese individuals in a Japanese population, as analyzed by terminal restriction fragment length polymorphism and next-generation sequencing. *BMC Gastroenterology*, *15*, 100. https://doi.org/10.1186/s12876-015-0330-2

37. Daisley, B. A., Koenig, D., Engelbrecht, K., Doney, L., Hards, K., Al, K. F., Reid, G., & Burton, J. P. (2021). Emerging connections between gut microbiome bioenergetics and chronic metabolic diseases. *Cell Reports*, *37*(10), 110087. https://doi.org/10.1016/j.celrep.2021.110087
38. Hu, J., Guo, P., Mao, R., Ren, Z., Wen, J., Yang, Q., Yan, T., Yu, J., Zhang, T., & Liu, Y. (2022). Gut microbiota signature of obese adults across different classifications. *Diabetes, Metabolic Syndrome and Obesity: Targets and Therapy*, *15*, 3933–3947. https://doi.org/10.2147/DMSO.S387523
39. Singh, P., Rawat, A., Alwakeel, M., Sharif, E., & Al Khodor, S. (2020). The potential role of vitamin D supplementation as a gut microbiota modifier in healthy individuals. *Scientific Reports*, *10*, article 21641. https://doi.org/10.1038/s41598-020-77806-4
40. Han, H., Yi, B., Zhong, R., Wang, M., Zhang, S., Ma, J., Yin, Y., Yin, J., Chen, L., & Zhang, H. (2021). From gut microbiota to host appetite: gut microbiota-derived metabolites as key regulators. *Microbiome*, *9*(1), 162. https://doi.org/10.1186/s40168-021-01093-y
41. Delzenne, N. M., & Cani, P. D. (2011). Gut microbiota and the pathogenesis of insulin resistance. *Current Diabetes Reports*, *11*(3), 154–159. https://doi.org/10.1007/s11892-011-0191-1
42. Sato, J., Kanazawa, A., Ikeda, F., Yoshihara, T., Goto, H., Abe, H., Komiya, K., Kawaguchi, M., Shimizu, T., Ogihara, T., Tamura, Y., Sakurai, Y., Yamamoto, R., Mita, T., Fujitani, Y., Fukuda, H., Nomoto, K., Takahashi, T., Asahara, T., . . . Watada, H. (2014). Gut dysbiosis and detection of "live gut bacteria" in blood of japanese patients with type 2 diabetes. *Diabetes Care*, *37*(8), 2343–2350.
43. Wu, H., Esteve, E., Tremaroli, V., Khan, M. T., Caesar, R., Mannerås-Holm, L., Ståhlman, M., Olsson, L. M., Serino, M., Planas-Félix, M., Xifra, G., Mercader, J. M., Torrents, D., Burcelin, R., Ricart, W., Perkins, R., Fernández-Real, J. M., & Bäckhed, F. (2017). Metformin alters the gut microbiome of individuals with treatment-naive type 2 diabetes, contributing to the therapeutic effects of the drug. *Nature Medicine*, *23*(7), 850–858. https://doi.org/10.1038/nm.4345
44. Yang, Y., Ren, R., Chen, Q., Zhang, Q., Wu, J., & Yin, D. (2022). *Coptis chinensis* polysaccharides dynamically influence the paracellular absorption pathway in the small intestine by modulating the intestinal mucosal immunity microenvironment. *Phytomedicine*, *104*, article 154322. https://doi.org/10.1016/j.phymed.2022.154322
45. Zhang, X.-Y., Chen, J., Yi, K., Peng, L., Xie, J., Gou, X., Peng, T., & Tang, L. (2020). *Gut Microbes*, *12*(1), article 1842990. https://doi.org/10.1080/19490976.2020.1842990
46. Nilsson, U., Rickard, Ö., Jägerstad, M., & Birkhed, D. (1988). Cereal fructans: *In vitro* and *in vivo* studies on availability in rats and humans. *The Journal of Nutrition*, *118*(11), 1325–1330.
47. Matheus, V. A., Monteiro, L., Oliveira, R. B., Maschio, D. A., & Collares-Buzato,

C. B. (2017). Butyrate reduces high-fat diet-induced metabolic alterations, hepatic steatosis and pancreatic beta cell and intestinal barrier dysfunctions in prediabetic mice. *Experimental Biology and Medicine*, *242*(12), 1214–1226. https://doi.org/10.1177/1535370217708188

48. Sjögren, K., Engdahl, C., Henning, P., Lerner, U. H., Tremaroli V., Lagerquist, M. K., Bäckhed, F., & Ohlsson, C. (2012). The gut microbiota regulates bone mass in mice. *Journal of Bone and Mineral Research*, *27*(6), 1357–1367. https://doi.org/10.1002/jbmr.1588
49. Xu, Z., Xie, Z., Sun, J., Huang, S., Chen, Y., Li, C., Sun, X., Xia, B., Tian, L., Guo, C., Li, F., & Pi, G. (2020). Gut microbiome reveals specific dysbiosis in primary osteoporosis. *Frontiers in Cellular and Infection Microbiology*, *10*, 160. https://doi.org/10.3389/fcimb.2020.00160
50. Das, M., Cronin, O., Keohane, D. M., Cormac, E. M., Nugent, H., Nugent, M., Molloy, C., O'Toole, P. W., Shanahan, F., Molloy, M. G., & Jeffery, I. B. (2019). Gut microbiota alterations associated with reduced bone mineral density in older adults. *Rheumatology*, *58*(12), 2295–2304. https://doi.org/10.1093/rheumatology/kez302
51. Sjögren, K., Engdahl, C., Henning, P., Lerner, U. H., Tremaroli, V., Lagerquist, M. K., Bäckhed, F., & Ohlsson, C. (2012). The gut microbiota regulates bone mass in mice. *Journal of Bone and Mineral Research*, *27*(6), 1357–1367. https://doi.org/10.1002/jbmr.1588
52. Cho, I., Yamanishi, S., Cox, L., Methé, B. A., Zavadil, J., Li, K., Gao, Z., Mahana, D., Raju, K., Teitler, I., Li, H., Alekseyenko, A. V., & Blaser, M. J. (2012). Antibiotics in early life alter the murine colonic microbiome and adiposity. *Nature*, *488*(7413), 621–626. https://doi.org/10.1038/nature11400
53. Pytlik, M., Folwarczna, J., & Janiec, W. (2004). Effects of doxycycline on mechanical properties of bones in rats with ovariectomy-induced osteopenia. *Calcified Tissue International*, *75*(3), 225–230. https://doi.org/10.1007/s00223-004-0097-x
54. Li, J.-Y., Chassaing, B., Tyagi, A. M., Vaccaro, C., Luo, T., Adams, J., Darby, T. M., Weitzmann, M. N., Mulle, J. G., Gewirtz, A. T., Jones, R. M., & Pacifici, R. (2016). Sex steroid deficiency–associated bone loss is microbiota dependent and prevented by probiotics. *The Journal of Clinical Investigation*, *126*(6), 2049–2063.
55. Carson, J. A., & Manolagas, S. C. (2015). Effects of sex steroids on bones and muscles: Similarities, parallels, and putative interactions in health and disease. *Bone*, *80*, 67–78.
56. Lucas, S., Omata, Y., Hofmann, J., Bottcher, M., Iljazovic, A., Sarter, K., Albrecht, O., Schulz, O., Krishnacoumar, B., Krönke, G., Herrmann, M., Mougiakakos, D., Strowig, T., Schett, G., & Zaiss, M. M. (2018). Short-chain fatty acids regulate systemic bone mass and protect from pathological bone loss. *Nature Communications*, *9*(1), 55. https://doi.org/10.1038/s41467-017-02490-4
57. Jansson, P.-A., Curiac, D., Ahrén, I. L., Hansson, F., Niskanen, T. M., Sjögren, K., & Ohlsson, C. (2019). Probiotic treatment using a mix of three *Lactobacillus* strains for lumbar spine bone loss in postmenopausal women: A randomised,

double-blind, placebo-controlled, multicentre trial. *The Lancet Rheumatology*, *1*(3), E154–E162. https://doi.org/10.1016/S2665-9913(19)30068-2

58. Dzutsev, A., Goldszmid, R. S., Viaud, S., Zitvogel, L., & Trinchieri, G. (2015). The role of the microbiota in inflammation, carcinogenesis, and cancer therapy. *European Journal of Immunology*, *45*(1), 17–31. https://doi.org/https://doi.org/10.1002/eji.20144497
59. Saffarian, A., Mulet, C., Regnault, B., Amiot, A., Tran-Van-Nhieu, J., Ravel, J., Sobhani, I., Sansonetti, P. J., & Pédron, T. (2019). Crypt- and mucosa-associated core microbiotas in humans and their alteration in colon cancer patients. *MBio*, *10*(4), e01315–e01319. https://doi.org/10.1128/mBio.01315-19
60. Wang, H., Hu, J., Wu, J., Ji, P., Shang, A., & Li, D. (2022). The function and molecular mechanism of commensal microbiome in promoting malignant progression of lung cancer. *Cancers*, *14*(21), 5394. https://doi.org/10.3390/cancers14215394
61. Guidi, R., Guerra, L., Levi, L., Stenerlöw, B., Fox, J. G., Josenhans, C., Masucci, M. G., & Frisan, T. (2013). Chronic exposure to the cytolethal distending toxins of gram-negative bacteria promotes genomic instability and altered DNA damage response. *Cellular Microbiology*, *15*(1), 98–113.
62. Castro-Mejía, J. L., Muhammed, M. K., Kot, W., Neve, H., Franz, C. M., Hansen, L. H., Vogensen, F. K., & Nielsen, D. S. (2015). Optimizing protocols for extraction of bacteriophages prior to metagenomic analyses of phage communities in the human gut. *Microbiome*, *3*, 64.
63. Wilson, M. R., Jiang, Y., Villalta, P. W., Stornetta, A., Boudreau, P. D., Carrá, A., Brennan, C. A., Chun, E., Ngo, L., Samson, L. D., Engelward, B. P., Garrett, W. S., Balbo, S., & Balskus, E. P. (2019). The human gut bacterial genotoxin colibactin alkylates DNA. *Science*, *363*(4628), eaar7785.
64. Chen, F., Zhuang, X., Lin, L., Yu, P., Wang, Y., Shi, Y., Hu, G., & Sun, Y. (2015). New horizons in tumor microenvironment biology: Challenges and opportunities. *BMC Medicine*, *13*(1), 45. https://doi.org/10.1186/s12916-015-0278-7
65. Gamallat, Y., Meyiah, A., Kuugbee, E. D ., Hago, A. M., Chiwala, G., Awadasseid, A., Bamba, D., Zhang, X., Shang, X., Luo, F., & Xin, Y. (2016). *Lactobacillus rhamnosus* induced epithelial cell apoptosis, ameliorates inflammation and prevents colon cancer development in an animal model. *Biomedicine & Pharmacotherapy*, *83*, 536–541. https://doi.org/10.1016/j.biopha.2016.07.001
66. Veziant, J., Gagnière, J., Jouberton, E., Bonnin, V., Sauvanet, P., Pezet, D., Barnich, N., Miot-Noirault, E., & Bonnet, M. (2016). Association of colorectal cancer with pathogenic *Escherichia coli*: Focus on mechanisms using optical imaging. *World Journal of Clinical Oncology*, 7(3), 293–301. https://doi.org/10.5306/wjco.v7.i3.293
67. Merali, Z. (2014, October 2). Physicists' model proposes evolutionary role for cancer. *Nature*. https://doi.org/10.1038/nature.2014.16068
68. Giovannini, C., Scazzocchio, B., Varì, R., Santangelo, C., D'Archivio, M., & Masella, R. (2007). Apoptosis in cancer and atherosclerosis: Polyphenol activities. *Annali dell'Istituto Superiore di Sanità*, *43*(4), 406–416.

69. Takashina, M., Inoue, S., Tomihara, K., Tomita, K., Hattori, K., Zhao, Q. L., Suzuki, T., Noguchi, M., Ohashi, W., & Hattori, Y. (2017). Different effect of resveratrol to induction of apoptosis depending on the type of human cancer cells. *International Journal of Oncology*, *50*(3), 787–797.
70. Kumar, S., Eroglu, E., Stokes, J. A. 3rd, Scissum-Gunn, K., Saldanha, S. N., Singh, U. P., Manne, U., Ponnazhagan, S., & Mishra, M. K. (2017). Resveratrol induces mitochondria-mediated, caspase-independent apoptosis in murine prostate cancer cells. *Oncotarget*, *8*(13), 20895–20908.
71. Wang, L., Jiang, G., Jing, N., Liu, X., Li, Q., Liang, W., & Liu, Z. (2020). Bilberry anthocyanin extracts enhance anti-PD-L1 efficiency by modulating gut microbiota. *Food & Function*, *11*(4), 3180–3190.
72. Griffin, L. E., Kohrt, S. E., Rathore, A., Kay, C. D., Grabowska, M. M., & Neilson, A. P. (2022). Microbial metabolites of flavanols in urine are associated with enhanced anti-proliferative activity in bladder cancer cells *in vitro*. *Nutrition and Cancer*, *74*(1), 194–210. https://doi.org/10.1080/01635581.2020.1869277
73. Kaźmierczak-Siedlecka, K., Marano, L., Merola, E., Roviello, F., & Połom, K. (2022). Sodium butyrate in both prevention and supportive treatment of colorectal cancer. *Frontiers in Cellular and Infection Microbiology*, *10*, article 2013806. https://doi.org/10.3389/fcimb.2022.1023806
74. Jin, Y., Dong, H., Xia, L., Yang, Y., Zhu, Y., Shen, Y., Zheng, H., Yao, C., Wang, Y., & Lu, S. (2019). The diversity of gut microbiome is associated with favorable responses to anti–programmed death 1 immunotherapy in Chinese patients with NSCLC. *Journal of Thoracic Oncology*, *14*(8), 1378–1389.
75. Matson, V., Fessler, J., Bao, R., Chongsuwat, T., Zha, Y., Alegre, M.-L., Luke, J. J., & Gajewski, T. F. (2018). The commensal microbiome is associated with anti-PD-1 efficacy in metastatic melanoma patients. *Science*, *359*(6371), 104–108.
76. Geller, L. T., Barzily-Rokni, M., Danino, T., Jonas, O. H., Shental, N., Nejman, D., Gavert, N., Zwang, Y., Cooper, Z. A., Shee, K., Thaiss, C. A., Reuben, A., Livny, J., Avraham, R., Frederick, D. T., Ligorio, M., Chatman, K., Johnston, S. E., Mosher, C. M., . . . Straussman, R. (2017). Potential role of intratumor bacteria in mediating tumor resistance to the chemotherapeutic drug gemcitabine. *Science*, *357*(6356), 1156–1160. https://doi.org/10.1126/science.aah5043
77. Al-Qadami, G., Van Sebille, Y., Le, H., & Bowen, J. (2019). Gut microbiota: Implications for radiotherapy response and radiotherapy-induced mucositis. *Expert Review of Gastroenterology & Hepatology*, *13*(5), 485–496. https://doi.org/10.1080/17474124.2019.1595586
78. Nejman, D., Livyatan, I., Fuks, G., Gavert, N., Zwang, Y., Geller, L. T., Rotter-Maskowitz, A., Weiser, R., Mallel, G., Gigi, E., Meltser, A., Douglaas, G. M., Kamer, I., Gopalakrishnan, V., Dadosh, T., Levin-Zaidman, S., Avnet, S., Atlan, T., Cooper, Z. A., . . . Straussman, R. (2020). The human tumor microbiome is composed of tumor type–specific intracellular bacteria. *Science*, *368*(6494), 973–980.
79. Dejea, C. M., Fathi, P., Craig, J. M., Boleij, A., Taddese, R., Geis, A. L., Wu, X., DeStefano Shields, C. E., Hechenbleikner, E. M., Huso, D. L., Anders, R. A., Gardielo, F. M., Wick, E. C., Want, H., Wu, S., Pardoll, D. M., Housseau, F., &

Sears, C. (2018). Patients with familial adenomatous polyposis harbor colonic biofilms containing tumorigenic bacteria. *Science, 359*(6375), 592–597.

80. Goodwin, A. C., DeStefano Shields, C. E., Wu, S., Huso, D. L., Wu, X., Murray-Stewart, T. R., Hacker-Prietz, A., Rabizadeh, S., Woster, P. M., Sears, C. L., & Casero, R. A. Jr. (2011). Polyamine catabolism contributes to enterotoxigenic *Bacteroides fragilis*–induced colon tumorigenesis. *Proceedings of the National Academy of Sciences of the United States of America, 108*(37), 15354–15359.
81. Bullman, S., Pedamallu, C. S., Sicinska, E., Clancy, T. E., Zhang, X., Cai, D., Neuberg, D., Huang, K., Guevara, F., Nelson, T., Chipashvili, O., Hagan, T., Walker, M., Ramachandran, A., Diosdado, B., Serna, G., Mulet, N., Landolfi, S., Ramón y Cajal, S., . . . Meyerson, M. (2017). Analysis of *Fusobacterium* persistence and antibiotic response in colorectal cancer. *Science, 358*(6369), 1443–1448.
82. Kostic, A. D., Chu, E., Robertson, L., Glickman, J. N., Gallini, C. A., Michaud, M., Clancy, T. E., Chung, D. C., Lochhead, P., Hold, G. L., El-Omar, E. M., Brenner, D., Fuchs, C. S., Meyerson, M., Garrett, W. S. (2013). *Fusobacterium nucleatum* potentiates intestinal tumorigenesis and modulates the tumor-immune microenvironment. *Cell Host & Microbe, 14*(2), 207–215.
83. Zhang, S., Yang, Y., Weng, W., Guo, B., Cai, G., Ma, Y., & Cai, S. (2019). *Fusobacterium nucleatum* promotes chemoresistance to 5-fluorouracil by upregulation of BIRC3 expression in colorectal cancer. *Journal of Experimental & Clinical Cancer Research, 38*(1), 14.
84. Jin, C., Lagoudas, G. K., Zhao, C., Bullman, S., Bhutkar, A., Hu, B., Ameh, S., Sandel, D., Liang, X. S., Mazzili, S., Whary, M. T., Meyerson, M., Germain, G. R., Blainey, P. C., Fox, J. G., & Jacks, T. (2019). Commensal microbiota promote lung cancer development via γδ T cells. *Cell, 176*(5), 998–1013.e16.
85. Bai, R., Lv, Z., Xu, D., & Cui, J. (2020). Predictive biomarkers for cancer immunotherapy with immune checkpoint inhibitors. *Biomarker Research, 8*, article 34.
86. Zheng, D.-W., Deng, W.-W., Song, W.-F., Wu, C.-C., Liu, J., Hong, S., Zhuang, Z.-N., Cheng, H., Sun, Z.-J., & Zhang, X.-Z. (2021). Biomaterial-mediated modulation of oral microbiota synergizes with PD-1 blockade in mice with oral squamous cell carcinoma. *Nature Biomedical Engineering, 6*(1), 32–43.
87. Ahn, J., Chen, C. Y., & Hayes, R. B. (2012). Oral microbiome and oral and gastrointestinal cancer risk. *Cancer Causes & Control, 23*(3), 399–404.
88. Han, Y. W., Shi, W., Huang, G. T.-J., Haake, S. K., Park, N.-H., Kuramitsu H., & Genco, R. J. (2000). Interactions between periodontal bacteria and human oral epithelial cells: *Fusobacterium nucleatum* adheres to and invades epithelial cells. *Infection and Immunity, 68*(6), 3140–3146.
89. Schmidt, B. L., Kuczynski, J., Bhattacharya, A., Huey, B., Corby, P. M., Queiroz, E.L.S., Nightingale, K., Kerr, A. R., DeLacure, M. D., Veeramachaneni, R., Olshen, A. B., & Albertson, D. G. (2014). Changes in abundance of oral microbiota associated with oral cancer. *PLOS ONE, 9*(6), e98741.
90. Irfan, M., Delgado, R.Z.R., & Frias-Lopez, J. (2020). The oral microbiome and cancer. *Frontiers in Immunology, 11*, article 591088.

91. Karpiński, T. M. (2019). Role of oral microbiota in cancer development. *Microorganisms*, 7(1), 20. https://doi.org/10.3390/microorganisms7010020
92. Garrett, W. S. (2019). The gut microbiota and colon cancer. *Science*, *364*(6446), 1133–1135.
93. Pushalkar, S., Hundeyin, M., Daley, D., Zambirinis, C. P., Kurz, E., Mishra, A., Mohan, N., Aykut, B., Usyk, M., Torres, L. E., Werba, G., Zhang, K., Guo, Y., Li, Q., Akkad, N., Lall, S., Wadowski, B., Gutierrez, J., Rossi, J.A.K., . . . Miller, G. (2018). The pancreatic cancer microbiome promotes oncogenesis by induction of innate and adaptive immune suppression. *Cancer Discovery*, *8*(4), 403–416.
94. Zhao, Y., Liu, Y., Li, S., Peng, Z., Liu, X., Chen, J., & Zheng, X. (2021). Role of lung and gut microbiota on lung cancer pathogenesis. *Journal of Cancer Research and Clinical Oncology*, *147*(8), 2177–2186. https://doi.org/10.1007/s00432-021-03644-0

CHAPTER 6: LEAKY GUT = LEAKY BRAIN

1. Plattner, V., Germann, B., Neuhaus, W., Noe, C., Gabor, F., & Wirth, M. (2010). Characterization of two blood-brain barrier mimicking cell lines: Distribution of lectin-binding sites and perspectives for drug delivery. *International Journal of Pharmaceutics*, *387*(1–2), 34–41. https://doi.org/10.1016/j.ijpharm.2009.11.030
2. Banks, W. A., Sharma, P., Bullock, K. M., Hansen, K. M., Ludwig, N., & Whiteside, T. L. (2020). Transport of extracellular vesicles across the blood-brain barrier: Brain pharmacokinetics and effects of inflammation. *International Journal of Molecular Sciences*, *21*(12), 4407. https://doi.org/10.3390/ijms21124407
3. Schepici, G., Silvestro, S., Bramanti, P., & Mazzon, E. (2019). The gut microbiota in multiple sclerosis: An overview of clinical trials. *Cell Transplantation*, *28*(12), 1507–1527. https://doi.org/10.1177/0963689719873890
4. Anand, N., Gorantla, V. R., & Chidambaram, S. B. (2023). The role of gut dysbiosis in the pathophysiology of neuropsychiatric disorders. *Cells*, *12*(1), 54. https://doi.org/10.3390/cells12010054
5. Obrenovich, M.E.M. (2018). Leaky gut, leaky brain? *Microorganisms*, *6*(4), 107. https://doi.org/10.3390/microorganisms6040107
6. Sun, Z., Song, Z.-G., Liu, C., Tan, S., Lin, S., Zhu, J., Dai, F.-H., Gao, J., She, J.-L., Mei, Z., Lou, T., Zheng, J.-J., Liu, Y., He, J., Zheng, Y., Ding, C., Qian, F., Zheng, Y., & Chen, Y.-M. (2022). Gut microbiome alterations and gut barrier dysfunction are associated with host immune homeostasis in COVID-19 patients. *BMC Medicine*, *20*(1), 24. https://doi.org/10.1186/s12916-021-02212-0
7. Powell, N., Walker, M. M., & Talley, N. J. (2017). The mucosal immune system: Master regulator of bidirectional gut-brain communications. *Nature Reviews Gastroenterology & Hepatology*, *14*(3), 143–159. https://doi.org/10.1038/nrgastro.2016.191
8. Forsythe, P., Bienenstock, J., & Kunze, W. A. (2014). Vagal pathways for microbiome-brain-gut axis communication. *Advances in Experimental Medicine and Biology*, *817*, 115–133. https://doi.org/10.1007/978-1-4939-0897-4_5

9. Tubbs, R. S., Rizk, E., Shoja, M. M., Loukas, M., Barbaro, N., & Spinner, R. J. (Eds.). (2015). *Nerves and nerve injuries* (Vol. 1). Academic Press.
10. Braniste, V., Al-Asmakh, M., Kowal, C., Anuar, F., Abbaspour, A., Tóth, M., Korecka, A., Bakocevic, N., Ng, L. G., Kundu, P., Gulyás, B., Halldin, C., Hultenby, K., Nilsson, H., Hebert, H., Volpe, B. T., Diamond, B., & Pettersson, S. (2014). The gut microbiota influences blood-brain barrier permeability in mice. *Science Translational Medicine*, *6*(263), 263ra158. https://doi.org/10.1126/scitranslmed.3009759. Erratum (2014) in: *Science Translational Medicine*, *6*(266), 266er7.
11. Soret, R., Chevalier, J., De Coppet, P., Poupeau, G., Derkinderen, P., Segain, J. P., & Neunlist, M. (2010). Short-chain fatty acids regulate the enteric neurons and control gastrointestinal motility in rats. *Gastroenterology*, *138*(5), 1772–1782.
12. Liu, H., Wang, J., He, T., Becker, S., Zhang, G., Li, D., & Ma, X. (2018). Butyrate: A double-edged sword for health? *Advances in Nutrition*, *9*(1), 21–29.
13. Resende, W. R., Valvassori, S. S., Réus, G. Z., Varela, R. B., Arent, C. O., Ribeiro, K. F., Bavaresco, D. V., Andersen, M. L., Zugno, A. I., & Quevedo, J. (2013). Effects of sodium butyrate in animal models of mania and depression: Implications as a new mood stabilizer. *Behavioural Pharmacology*, *24*(7), 569–579.
14. Valvassori, S. S., Resende, W. R., Budni, J., Dal-Pont, G. C., Bavaresco, D. V., Réus, G. Z., Carvalho, A. F., Conçalves, C. L., Furlanetto, C. B., Streck, E. L., & Quevedo, J. (2015). Sodium butyrate, a histone deacetylase inhibitor, reverses behavioral and mitochondrial alterations in animal models of depression induced by early- or late-life stress. *Current Neurovascular Research*, *12*(4), 312–320.
15. Gao, K., Pi, Y., Mu, C.-L., Peng, Y., Huang, Z., & Zhu, W.-Y. (2018). Antibiotics-induced modulation of large intestinal microbiota altered aromatic amino acid profile and expression of neurotransmitters in the hypothalamus of piglets. *Journal of Neurochemistry*, *146*(3), 219–234. https://doi.org/10.1111/jnc.14333
16. Caspani, G., & Swann, J. (2019). Small talk: Microbial metabolites involved in the signaling from microbiota to brain. *Current Opinion in Pharmacology*, *48*, 99–106. https://doi.org/10.1016/j.coph.2019.08.001
17. Wikoff, W. R., Anfora, A. T., Liu, J., Schultz, P. G., Lesley, S. A., Peters, E. C., & Siuzdak, G. (2009). Metabolomics analysis reveals large effects of gut microflora on mammalian blood metabolites. *Proceedings of the National Academy of Sciences of the United States of America*, *106*(10), 3698–3703. https://doi.org/10.1073/pnas.0812874106
18. Matsumoto, M., Ooga, T., Kibe, R., Aiba, Y., Koga, Y., & Benno, Y. (2017). Colonic absorption of low-molecular-weight metabolites influenced by the intestinal microbiome: A pilot study. *PLOS ONE*, *12*(1), e0169207. https://doi.org/10.1371/journal.pone.0169207
19. Gao, K., Pi, Y., Mu, C.-L., Peng, Y., Huang, Z., & Zhu, W.-Y. (2018). Antibiotics-induced modulation of large intestinal microbiota altered aromatic amino acid profile and expression of neurotransmitters in the hypothalamus

of piglets. *Journal of Neurochemistry*, *146*(3), 219–234. https://doi.org/10.1111/jnc.14333

20. Fujisaka, S., Avila-Pacheco, J., Soto, M., Kostic, A., Dreyfuss, J. M., Pan, H., Ussar, S., Altindis, E., Li, N., Bry, L., Clish, C. B., & Kahn, C. R. (2018). Diet, genetics, and the gut microbiome drive dynamic changes in plasma metabolites. *Cell Reports*, *22*(11), 3072–3086. https://doi.org/10.1016/j.celrep.2018.02.060
21. Frost, G., Sleeth, M. L., Sahuri-Arisoylu, M., Lizarbe, B., Cerdan, S., Brody, L., Anastasovska, J., Ghourab, S., Hankir, M., Zhang, S., Carling, D., Swann, J. R., Gibson, G., Viardot, A., Morrison, D., Thomas, E. L., & Bell, J. E. (2014). The short-chain fatty acid acetate reduces appetite via a central homeostatic mechanism. *Nature Communications*, *5*, 3611. https://doi.org/10.1038/ncomms4611
22. Kaelberer, M. M., Rupprecht, L. E., Liu, W. W., Weng, P., & Bohórquez, D. V. (2020). Neuropod cells: The emerging biology of gut-brain sensory transduction. *Annual Review of Neuroscience*, *43*, 337–353. https://doi.org/10.1146/annurev-neuro-091619-022657
23. Kaelberer, M. M., Buchanan, K. L., Klein, M. E., Barth, B. B., Montoya, M. M., Shen, X., & Bohórquez, D. V. (2018). A gut-brain neural circuit for nutrient sensory transduction. *Science*, *361*(6408), eaat5236. https://doi.org/10.1126/science.aat5236
24. Checa-Ros, A., Jeréz-Calero, A., Molina-Carballo, A., Campoy, C., & Muñoz-Hoyos, A. (2021). Current evidence on the role of the gut microbiome in ADHD pathophysiology and therapeutic implications. *Nutrients*, *13*(1), 249. https://doi.org/10.3390/nu13010249
25. Bruckner, J. J., Stednitz, S. J., Grice, M. Z., Zaidan, D., Massaquoi, M. S., Larsch, J., Tallafuss, A., Guillemin, K., Washbourne, P., & Eisen, J. S. (2022). The microbiota promotes social behavior by modulating microglial remodeling of forebrain neurons. *PLOS Biology*, *20*(11), e3001838. https://doi.org/10.1371/journal.pbio.3001838
26. Li, Q., & Barres, B. A. (2018). Microglia and macrophages in brain homeostasis and disease. *Nature Reviews Immunology*, *18*(4), 225–242.
27. Zhan, Y., Paolicelli, R. C., Sforazzini, F., Weinhard, L., Bolasco, G., Pagani, F., Vyssotski, A. L., Bifone, A., Gozzi, A., Ragozzino, D., & Gross, C. T. (2014). Deficient neuron-microglia signaling results in impaired functional brain connectivity and social behavior. *Nature Neuroscience*, *17*(3), 400–406.
28. Smith, C. J. (2021). Emerging roles for microglia and microbiota in the development of social circuits. *Brain, Behavior, & Immunity—Health*, *16*, 100296.
29. Butler, C. A., Popescu, A. S., Kitchener, E.J.A., Allendorf, D. H., Puigdellívol, M., & Brown, G. C. (2021). Microglial phagocytosis of neurons in neurodegeneration, and its regulation. *Journal of Neurochemistry*, *158*(3), 621–639. https://doi.org/10.1111/jnc.15327
30. Cooke, M. B., Catchlove, S., & Tooley, K. L. (2022). Examining the influence of the human gut microbiota on cognition and stress: A systematic review of the literature. *Nutrients*, *14*(21), 4623. https://doi.org/10.3390/nu14214623
31. Meldrum, B. S. (2000). Glutamate as a neurotransmitter in the brain: Review of physiology and pathology. *The Journal of Nutrition*, *130*(4)Suppl.), S1007–S1015. https://doi.org/10.1093/jn/130.4.1007S

32. Brekke, E., Morken, T. S., Walls, A. B., Waagepetersen, H., Schousboe, A., & Sonnewald, U. (2016). Anaplerosis for glutamate synthesis in the neonate and in adulthood. *Advances in Neurobiology*, *13*, 43–58. https://doi.org/10.1007/978-3-319-45096-4_3
33. Kaelberer, M. M., Buchanan, K. L., Klein, M. E., Barth, B. B., Montoya, M. M., Shen, X., & Bohórquez, D. V. (2018). A gut-brain neural circuit for nutrient sensory transduction. *Science*, *361*(4608), eaat5236. https://doi.org/10.1126/science.aat5236
34. Mitani, H., Shirayama, Y., Yamada, T., Maeda, K., Ashby, C. R. Jr., & Kawahara, R. (2006). Correlation between plasma levels of glutamate, alanine and serine with severity of depression. *Progress in Neuro-psychopharmacology & Biological Psychiatry*, *30*(6), 1155–1158.
35. Holemans, S., De Paermentier, F., Horton, R. W., Crompton, M. R., Katona, C. L., & Maloteaux, J. M. (1993). NMDA glutamatergic receptors, labelled with [3H]MK-801, in brain samples from drug-free depressed suicides. *Brain Research*, *616*(1–2), 138–143.
36. Frye, M. A., Tsai, G. E, Huggins, T., Coyle, J. T., & Post, R. M. (2007). Low cerebrospinal fluid glutamate and glycine in refractory affective disorder. *Biological Psychiatry*, *61*(2), 162–166.
37. Lydiard, R. B. (2003). The role of GABA in anxiety disorders. *The Journal of Clinical Psychiatry*, *64*, 21–27.
38. Lee, S.-E., Lee, Y., & Lee, G. H. (2019). The regulation of glutamic acid decarboxylases in GABA neurotransmission in the brain. *Archives of Pharmacal Research*, *42*(12), 1031–1039. https://doi.org/10.1007/s12272-019-01196-z
39. Frost, G., Sleeth, M. L., Sahuri-Arisoylu, M., Lizarbe, B., Cerdan, S., Brody, L., Anastasovska, J., Ghourab, S., Hankir, M., Zhang, S., Carling, D., Swann, J. R., Gibson, G., Viardot, A., Morrison, D., Thomas, E. L., & Bell, J. D. (2014). The short-chain fatty acid acetate reduces appetite via a central homeostatic mechanism. *Nature Communications*, *5*, 3611. https://doi.org/10.1038/ncomms4611
40. Picciotto, M. R., Higley, M. J., & Mineur, Y. S. (2012). Acetylcholine as a neuromodulator: Cholinergic signaling shapes nervous system function and behavior. *Neuron*, *76*(1), 116–129. https://doi.org/10.1016/j.neuron.2012.08.036
41. Koussoulas, K., Swaminathan, M., Fung, C., Bornstein, J. C., & Foong, J.P.P. (2018). Neurally released GABA acts via GABAC receptors to modulate $Ca^{2}+$ transients evoked by trains of synaptic Inputs, but not responses evoked by single stimuli, in myenteric neurons of mouse ileum. *Frontiers in Physiology*, *9*, 97. https://doi.org/10.3389/fphys.2018.00097
42. Horiuchi, Y., Kimura, R., Kato, N., Fujii, T., Seki, M., Endo, T., Kato, T., & Kawashima, K. (2003). Evolutional study on acetylcholine expression. *Life Sciences*, *72*(15), 1745–1756. https://doi.org/10.1016/S0024-3205(02)02478-5
43. Amenta, F., & Tayebati, S. K. (2008). Pathways of acetylcholine synthesis, transport and release as targets for treatment of adult-onset cognitive dysfunction. *Current Medicinal Chemistry*, *15*(5), 488–498. https://doi.org/10.2174/092986708783503203
44. Ferreira-Vieira, T. H., Guimaraes, I. M., Silva, F. R., & Ribeiro, F. M. (2016).

Alzheimer's disease: Targeting the cholinergic system. *Current Neuropharmacology*, *14*(1), 101–115. https://doi.org/10.2174/1570159X13666150716165726

45. Wang, C., Zheng, D., Weng, F., Jin, Y., & He, L. (2022). Sodium butyrate ameliorates the cognitive impairment of Alzheimer's disease by regulating the metabolism of astrocytes. *Psychopharmacology*, *239*(1), 215–227. https://doi.org/10.1007/s00213-021-06025-0
46. Binosha Fernando, W.M.A.D., Martins, I. J., Morici, M., Bharadwaj, P., Rainey-Smith, S. R., Lim, W.L.F., & Martins, R. N. (2020). Sodium butyrate reduces brain amyloid-β bevels and improves cognitive memory performance in an Alzheimer's disease transgenic mouse model at an early disease stage. *Journal of Alzheimer's Disease*, *74*(1), 91–99. https://doi.org/10.3233/JAD-190120
47. Eisenhofer, G., Aneman, A., Friberg, P., Hooper, D., Fandriks, L., Lonroth, H., Hunyady, B., & Mezey, E. (1997). Substantial production of dopamine in the human gastrointestinal tract. *The Journal of Clinical Endocrinology & Metabolism*, *82*(11), 3864–3871. https://doi.org/10.1210/jcem.82.11.4339
48. Meyer, J. H., Krüger, S., Wilson, A. A., Christensen, B. K., Goulding, V. S., Schaffer, A., Minifie, C., Houle, S., Hussey, D., & Kennedy, S. (2001). Lower dopamine transporter binding potential in striatum during depression. *NeuroReport*, *12*(18), 4121–4125. https://doi.org/10.1097/00001756-200112210-00052
49. Vaughan, C. J., Aherne, A. M., Lane, E., Power, O., Carey, R. M., & O'Connell, D. P. (2000). Identification and regional distribution of the dopamine D(1A) receptor in the gastrointestinal tract. *American Journal of Physiology: Regulatory, Integrative and Comparative Physiology*, *279*(2), R599–R609. https://doi.org/10.1152/ajpregu.2000.279.2.R599
50. Gershon, M. D. (2013). 5-Hydroxytryptamine (serotonin) in the gastrointestinal tract. *Current Opinion in Endocrinology & Diabetes*, *20*(1), 14–21.
51. Helton, S. G., & Lohoff, F. W. (2015). Serotonin pathway polymorphisms and the treatment of major depressive disorder and anxiety disorders. *Pharmacogenomics*, *16*(5), 541–553. https://doi.org/10.2217/pgs.15.15
52. Booij, L., Van der Does, W., Benkelfat, C., Bremner, J. D., Cowen, P. J., Fava, M., Gillin, C., Leyton, M., Moore, P., Smith, K. A., & Van der Kloot, W. A. (2002). Predictors of mood response to acute tryptophan depletion: A reanalysis. *Neuropsychopharmacology*, *27*(5), 852–861. https://doi.org/10.1016/S0893-133X(02)00361-5
53. Yano, J. M., Yu, K., Donaldson, G. P., Shastri, G. G., Ann, P., Ma, L., Nagler, C. R., Ismagilov, R. F., Mazmanian, S. K., & Hsiao, E. Y. (2015). Indigenous bacteria from the gut microbiota regulate host serotonin biosynthesis. *Cell*, *161*(2), 264–276. https://doi.org/10.1016/j.cell.2015.02.047
54. Glavin, G. B., & Szabo, S. (1990). Dopamine in gastrointestinal disease. *Digestive Diseases and Sciences*, *35*(9), 1153–1161. https://doi.org/10.1007/BF01537589
55. Luqman, A., Nega, M., Nguyen, M.-T., Ebner, P., & Gotz, F. (2018). SadA-expressing staphylococci in the human gut show increased cell adherence and internalization. *Cell Reports*, *22*(2), 535–545. https://doi.org/10.1016/j.celrep.2017.12.058

56. Barnett, J. A., & Gibson, D. L. (2020). Separating the empirical wheat from the pseudoscientific chaff: A critical review of the literature surrounding glyphosate, dysbiosis and wheat-sensitivity. *Frontiers in Microbiology*, *11*, article 556729. https://doi.org/10.3389/fmicb.2020.556729
57. Winter, G., Hart, R. A., Charlesworth, R.P.G., & Sharpley, C. F. (2018). Gut microbiome and depression: What we know and what we need to know. *Reviews in the Neurosciences*, *29*(6), 629–643.
58. Liu, L., Wang, H., Zhang, H., Chen, X., Zhang, Y., Wu, J., Zhao, L., Wang, D., Pu, J., Ji, P., & Xie, P. (2022). Toward a deeper understanding of gut microbiome in depression: The promise of clinical applicability. *Advanced Science*, *9*(35), article 202203707. https://doi.org/10.1002/advs.202203707
59. Liu, L., Wang, H., Zhang, H., Chen, X., Zhang, Y., Wu, J., Zhao, L., Wang, D., Pu, J., Ji, P., & Xie, P. (2022). Toward a deeper understanding of gut microbiome in depression: The promise of clinical applicability. *Advanced Science*, *9*(35), article 202203707. https://doi.org/10.1002/advs.202203707
60. Jiang, H.-Y., Zhang, X., Yu, Z.-H., Zhang, Z., Deng, M., Zhao, J.-H., & Ruan, B. (2018). Altered gut microbiota profile in patients with generalized anxiety disorder. *Journal of Psychiatric Research*, *104*, 130–136. https://doi.org/10.1016/j.jpsychires.2018.07.007
61. Zheng, P., Zeng, B., Zhou, C., Liu, M., Fang, Z., Xu, X., Zeng, L., Chen, J., Fan, S., Du, X., Zhang, X., Yang, D., Yang, Y., Meng, H., Li, W., Melgiri, N. D., Licinio, J., Wei, H., & Xie, P. (2016). Gut microbiome remodeling induces depressive-like behaviors through a pathway mediated by the host's metabolism. *Molecular Psychiatry*, *21*(6), 786–796. https://doi.org/10.1038/mp.2016.44
62. Radjabzadeh, D., Bosch, J. A., Uitterlinden, A. G., Zwinderman, A. H., Ikram, M. A., van Meurs, J.B.J., Luik, A. I., Nieuwdorp, M., Lok, A., van Duijn, C. M., Kraaij, R., & Amin, N. (2022). Gut microbiome-wide association study of depressive symptoms. *Nature Communications*, *13*(1), 7128. https://doi.org/10.1038/s41467-022-34502-3
63. Safadi, J. M., Quinton, A.M.G., Lennox, B. R., Burnet, P.W.J., & Minichino, A. (2022). Gut dysbiosis in severe mental illness and chronic fatigue: a novel trans-diagnostic construct? A systematic review and meta-analysis. *Molecular Psychiatry*, *27*(1), 141–153. https://doi.org/10.1038/s41380-021-01032-1
64. Stevens, B. R., Goel, R., Seungbum, K., Richards, E. M., Holbert, R. C., Pepine, C. J., & Raizada, M. K. (2018). Increased human intestinal barrier permeability plasma biomarkers zonulin and FABP2 correlated with plasma LPS and altered gut microbiome in anxiety or depression. *Gut*, *67*(8), 1555–1557. https://doi.org/10.1136/gutjnl-2017-314759
65. Navarro-Tapia, E., Almeida-Toledano, L., Sebastiani, G., Serra-Delgado, M., García-Algar, Ó., & Andreu-Fernández, V. (2021). Effects of microbiota imbalance in anxiety and eating disorders: Probiotics as novel therapeutic approaches. *International Journal of Molecular Sciences*. *22*(5), 2351. https://doi.org/10.3390/ijms22052351
66. Yang, Y.-J., Chen, C.-N., Zhan, J.-Q., Liu, Q.-S., Liu, Y., Jiang, S.-Z., & Wei, B. (2021). Decreased plasma hydrogen sulfide level is associated with the

severity of depression in patients with depressive disorder. *Frontiers in Psychiatry*, *12*, article 765664. https://doi.org/10.3389/fpsyt.2021.765664
67. Zimmermann, M., Zimmermann-Kogadeeva, M., Wegmann, R., & Goodman, A. (2019). Mapping human microbiome drug metabolism by gut bacteria and their genes. *Nature*, *570*(7762), 462–467.
68. Shen, Y., Yang, X., Li, G., Gao, J., & Liang, Y. (2021). The change of gut microbiota in MDD patients under SSRIs treatment. *Scientific Reports*, *11*, article 14918. https://doi.org/10.1038/s41598-021-94481-1
69. Lyte, M., & Brown, D. R. (2018). Evidence for PMAT- and OCT-like biogenic amine transporters in a probiotic strain of *Lactobacillus*: Implications for interkingdom communication within the microbiota-gut-brain axis. *PLOS ONE*, *13*(1), e0191037.
70. Li, B., Xu, M., Wang, Y., Feng, L., Xing, H., & Zhang, K. (2023). Gut microbiota: A new target for traditional Chinese medicine in the treatment of depression. *Journal of Ethnopharmacology*, *303*, article 116038. https://doi.org/10.1016/j.jep.2022.116038
71. Warnecke, T., Schäfer, K. H., Claus, I., Del Tredici, K., & Jost, W. H. (2022). Gastrointestinal involvement in Parkinson's disease: Pathophysiology, diagnosis, and management. *NPJ Parkinson's Disease*, *8*(1), 31.
72. Rolli-Derkinderen, M., Leclair-Visonneau, L., Bourreille, A., Coron, E., Neunlist, M., & Derkinderen, P. (2019). Is Parkinson's disease a chronic low-grade inflammatory bowel disease? *Journal of Neurology*, *267*(8), 2207–2213.
73. Sun, M.-F., & Shen, Y.-Q. (2018). Dysbiosis of gut microbiota and microbial metabolites in Parkinson's disease. *Ageing Research Reviews*, *45*, 53–61. https://doi.org/10.1016/j.arr.2018.04.004
74. Baert, F., Matthys, C., Maselyne, J., Van Poucke, C., Van Coillie, E., Bergmans, B., & Vlaemynck, G. (2021). Parkinson's disease patients' short chain fatty acids production capacity after *in vitro* fecal fiber fermentation. *NPJ Parkinson's Disease*, *7*, article 72. https://doi.org/10.1038/s41531-021-00215-5
75. Aho, V.T.E., Houser, M. C., Pereira, P.A.B., Chang, J., Rudi, K., Paulin, L., Hertzberg, V., Auvinen, P., Tansey, M. G., & Scheperjans, F. (2021). Relationships of gut microbiota, short-chain fatty acids, inflammation, and the gut barrier in Parkinson's disease. *Molecular Neurodegeneration*, *16*(1), 6. https://doi.org/10.1186/s13024-021-00427-6
76. Yang, X., Ai, P., He, X., Mo, C., Zhang, Y., Xu, S., Lai, Y., Qian, Y., & Xiao, Q. (2022). Parkinson's disease is associated with impaired gut-blood barrier for short-chain fatty acids. *Movement Disorders*, *37*(8), 1634–1643.
77. Chen, S.-J., Chi, Y.-C., Ho, C.-H., Yang, W.-S., & Lin, C.-H. (2021). Plasma lipopolysaccharide-binding protein reflects risk and progression of Parkinson's disease. *Journal of Parkinson's Disease*, *11*(3), 1129–1139.
78. Selkoe, D. J. (2003). Folding proteins in fatal ways. *Nature*, *426*(6968), 900–904.
79. Sitia, R., & Braakman, I. (2003). Quality control in the endoplasmic reticulum protein factory. *Nature*, *426*(6968), 891–894.
80. Taylor, J. P., Hardy, J., & Fischbeck, K. H. (2002). Toxic proteins in neurodegenerative disease. *Science*, *296*(5575), 1991–1995.

81. Kalia, L. V., & Lang, A. E. (2015). Parkinson's disease. *The Lancet*, *386*(9996), 896–912.
82. Goedert, M., Spillantini, M. G., Del Tredici, K., & Braak, H. (2013). 100 years of Lewy pathology. *Nature Reviews Neurology*, *9*(1), 13–24.
83. Wang, C., Lau, C. Y., Ma, F., & Zheng, C. (2021). Genome-wide screen identifies curli amyloid fibril as a bacterial component promoting host neurodegeneration. *Proceedings of the National Academy of Sciences of the United States of America*, *118*(34), e2106504118. https://doi.org/10.1073/pnas.2106504118
84. Friedland, R. P., & Chapman, M. R. (2017). The role of microbial amyloid in neurodegeneration. *PLOS Pathogens*, *13*(12), e1006654. https://doi.org/10.1371/journal.ppat.1006654
85. Walker, A. C., Bhargava, R., Vaziriyan-Sani, A. S., Pourciau, C., Donahue, E. T., Dove, A. S., Gebhardt, M. J., Ellward, G. L., Romeo, T., & Czyż, D. M. (2021). Colonization of the *Caenorhabditis elegans* gut with human enteric bacterial pathogens leads to proteostasis disruption that is rescued by butyrate. *PLOS Pathogens*, *17*, e1009510. https://doi.org/10.1371/journal.ppat.1009510
86. Cherny, I., Rockah, L., Levy-Nissenbaum, O., Gophna, U., Ron, E. Z., & Gazit, E. (2005). The formation of *Escherichia coli* Curli amyloid fibrils is mediated by prion-like peptide repeats. *Journal of Molecular Biology*, *352*(2), 245–252. https://doi.org/10.1016/j.jmb.2005.07.028
87. Friedland, R. P., & Chapman, M. R. (2017). The role of microbial amyloid in neurodegeneration. *PLOS Pathogens*, *13*(12), e1006654. https://doi.org/10.1371/journal.ppat.1006654
88. Zaborina, O., Kohler, J. E., Wang, Y., Bethel, C., Shevchenko, O., Wu, L., Turner, J. R., & Alverdy, J. C. (2006). Identification of multi-drug resistant *Pseudomonas aeruginosa* clinical isolates that are highly disruptive to the intestinal epithelial barrier. *Annals of Clinical Microbiology and Antimicrobials*, *5*, 14. https://doi.org/10.1186/1476-0711-5-14
89. Voth, S., Gwin, M., Francis, C. M., Balczon, R., Frank, D. W., Pittet, J.-F., Wagener, B. M., Moser, S. A., Alexeyev, M., Housley, N., Audia, J. P., Piechocki, S., Madera, K., Simmons, A., Crawford, M., & Stevens, T. (2020). Virulent *Pseudomonas aeruginosa* infection converts antimicrobial amyloids into cytotoxic prions. *The FASEB Journal*, *34*(7), 9156–9179. https://doi.org/10.1096/fj.202000051RRR
90. Balczon, R., Morrow, K. A., Zhou, C., Edmonds, B., Alexeyev, M., Pittet, J.-F., Wagener, B. M., Moser, S. A., Leavesley, S., Zha, X., Frank, D. W., & Stevens, T. (2017). Pseudomonas *aeruginosa* infection liberates transmissible, cytotoxic prion amyloids. *The FASEB Journal*, *31*(7), 2785–2796. https://doi.org/10.1096/fj.201601042RR
91. Murros, K. E., Huynh, V. A., Takala, T. M., & Saris, P.E.J. (2021). *Desulfovibrio* bacteria are associated with Parkinson's disease. *Frontiers in Cellular and Infection Microbiology*, *11*, article 652617. https://doi.org/10.3389/fcimb.2021.652617
92. Holmqvist, S., Chutna, O., Bousset, L., Aldrin-Kirk, P., Li, W., Björklund, T., Wang, Z.-Y., Roybon, L., Melki, R., & Li, J.-Y. (2014). Direct evidence of

Parkinson pathology spread from the gastrointestinal tract to the brain in rats. *Acta Neuropathologica*, *128*(6), 805–820.
93. Zhao, Y., Dua, P., & Lukiw, W. J. (2015). Microbial sources of amyloid and relevance to amyloidogenesis and Alzheimer's disease (AD). *Journal of Alzheimer's Disease & Parkinsonism*, *5*, 177.
94. Friedland, R. P. (2015). Mechanisms of molecular mimicry involving the microbiota in neurodegeneration. *Journal of Alzheimer's Disease*, *45*(2), 349–362. https://doi.org/10.3233/JAD-142841
95. Friedland, R. P. (2015). Mechanisms of molecular mimicry involving the microbiota in neurodegeneration. *Journal of Alzheimer's Disease*, *45*(2), 349–362. https://doi.org/10.3233/JAD-142841
96. Bunyoz, A. H., Christensen, R.H.B., Orlovska-Waast, S., Nordentoft, M., Mortensen, P. B., Petersen, L. V., & Benros, M. E. (2022). Vagotomy and the risk of mental disorders: A nationwide population-based study. *Acta Psychiatrica Scandinavica*, *145*(1), 67–78. https://doi.org/10.1111/acps.13343
97. Pan-Montojo, F., Schwarz, M., Winkler, C., Arnhold, M., O'Sullivan, G. A., Pal, A., Said, J., Marsico, G., Verbavatz, J.-M., Rodrigo-Angulo, M., Gille, G., Funk, R.H.W., & Reichmann, H. (2012). Environmental toxins trigger PD-like progression via increased alpha-synuclein release from enteric neurons in mice. *Scientific Reports*, *2*(1), 898.
98. Kim, S., Kwon, S.-H., Kam, T.-I., Panicker, N., Karuppagounder, S. S., Lee, S., Lee, J. H., Kim, W. R., Kook, M., Foss, C. A., Shen, C., Lee, H., Kulkami, S., Pasricha, P. J., Lee, G., Pomper, M. G., Dawson, V. L., Dawson, T. M., & Ko, H. S. (2019). Transneuronal propagation of pathologic α-synuclein from the gut to the brain models Parkinson's disease. *Neuron*, *103*(4), 627–641.e7.
99. Kumar, D.K.V., Choi, S. H., Washicosky, K. J., Eimer, W. A., Tucker, S., Ghofrani, J., Lefkowitz, A., McColl, G., Goldstein, L. E., Tanzi, R. E., & Moir, R. D. (2016). Amyloid-β peptide protects against microbial infection in mouse and worm models of Alzheimer's disease. *Science Translational Medicine*, *8*(340), article 340ra72. https://doi.org/10.1126/scitranslmed.aaf1059
100. Zhao, Y., Jaber, V., & Lukiw, W. J. (2017). Secretory products of the human GI tract microbiome and their potential impact on Alzheimer's disease (AD): Detection of lipopolysaccharide (LPS) in AD hippocampus. *Frontiers in Cellular and Infection Microbiology*, *7*, 318. https://doi.org/10.3389/fcimb.2017.00318
101. Zhao, Y., Dua, P., & Lukiw, W. J. (2015). Microbial sources of amyloid and relevance to amyloidogenesis and Alzheimer's disease (AD). *Journal of Alzheimer's Disease & Parkinsonism*, *5*, 177.
102. Paasila, P. J., Aramideh, J. A., Sutherland, G. T., & Graeber, M. B. (2022). Synapses, microglia, and lipids in Alzheimer's disease. *Frontiers in Neuroscience*, *15*, article 778822. https://doi.org/10.3389/fnins.2021.778822
103. Kesika, P., Suganthy, N., Sivamaruthi, B. S., & Chaiyasut, C. (2021). Role of gut-brain axis, gut microbial composition, and probiotic intervention in Alzheimer's disease, *Life Sciences*, *264*, article 118627. https://doi.org/10.1016/j.lfs.2020.118627

104. Friedland, R. P. (2015). Mechanisms of molecular mimicry involving the microbiota in neurodegeneration. *Journal of Alzheimer's Disease*, *45*(2), 349–362. https://doi.org/10.3233/JAD-142841
105. Jain, T., & Li, Y.-M. (2023). Gut microbes modulate neurodegeneration, *Science*, *379*(6628), 142–143. https://doi.org10.1126/science.adf9548
106. Bozelli, J. C. Jr., Azher, S., & Epand, R. M. (2021). Plasmalogens and chronic inflammatory diseases. *Frontiers in Psychiatry*, *12*, article 730829. https://doi.org/10.3389/fphys.2021.730829
107. Bizeau, J.-B., Albouery, M., Grégoire, S., Buteau, B., Martine, L., Crépin, M., Bron, A. M., Berdeaux, O., Acar, N., Chassaing, B., & Bringer, M.-A. (2022). Dietary inulin supplementation affects specific plasmalogen species in the brain. *Nutrients*, *14*(15), 3097.
108. Kaiser, J. (2021, October 7). The most common Alzheimer's risk gene may also protect against memory loss. *Science*. https://www.science.org/content/article/most-common-alzheimer-s-risk-gene-may-also-protect-against-memory--loss.
109. Patrick, R. P. (2019). Role of phosphatidylcholine-DHA in preventing APOE4-associated Alzheimer's disease. *The FASEB Journal*, *33*(2), 1554–1564. doi:10.1096/fj.201801412R
110. Calder, P. C. (2016). The DHA content of a cell membrane can have a significant influence on cellular behaviour and responsiveness to signals. *Annals of Nutrition & Metabolism*, *69*, 8.
111. Barberger-Gateau, P., Samieri, C., Féart, C., & Plourde, M. (2011). Dietary omega 3 polyunsaturated fatty acids and Alzheimer's disease: Interaction with apolipoprotein E genotype. *Current Alzheimer Research*, *8*(5), 479–491. https://doi.org/10.2174/156720511796391926
112. Qin, Y., Havulinna, A. S., Liu, Y., Jousilahti, P., Ritchie, S. C., Tokolyi, A., Sanders, J. G., Valsta, L., Brozyńska, M., Zhu, Q., Tripathi, A., Vázquez-Baeza, Y., Loomba, R., Cheng, S., Jain, M., Niiranen, T., Lahti, L., Knight, R., Salomaa, V., . . . Guillaume, M. (2022). Combined effects of host genetics and diet on human gut microbiota and incident disease in a single population cohort. *Nature Genetics*, *54*(5), 134–142. https://doi.org/10.1038/s41588-021-00991-z
113. Grieneisen, L., Dausani, M., Gould, T., Björk, J. R., Grenier, J.-C., Yotova, V., Jansen, D., Gottel, N., Gordon, J. B., Learn, N. H., Gesquiere, L. R., Wango, T. L., Mututua, R. S., Warutere, J. K., Siodi, L., Gilbert, J. A., Barreiro, L. B., Alberts, S. C., Tung, J., . . . Blekhman, R. (2021). Gut microbiome heritability is nearly universal but environmentally contingent. *Science*, *373*(6551), 181–186.
114. Tran, T.T.T., Corsini, S., Kellingray, L., Hegarty, C., Le Gall, G., Narbad, A., Müller, M., Tejera, N., O'Toole, P. W., Minihane, A.-M., & Vauzour, D. (2019). *APOE* genotype influences the gut microbiome structure and function in humans and mice: Relevance for Alzheimer's disease pathophysiology. *The FASEB Journal*, *33*(7), 8221–8231. https://doi.org/10.1096/fj.201900071R
115. Seo, D.-O., O'Donnell, D., Jain, N., Urich, J. D., Herz, J., Li, Y., Lemieux, M., Cheng, J., Hu, H., Serrano, J. R., Bao, X., Franke, E., Karlsson, M., Meier, M., Deng, S., Desai, C., Dodiya, H., Lelwala-Guruge, J., Handley, S. A., . . .

Holtzman, D. M. (2023). ApoE isoform- and microbiota-dependent progression of neurodegeneration in a mouse model of tauopathy. *Science*, *379*(7), eadd1236.

116. Nichols, R. G., & Davenport, E. R. (2021). The relationship between the gut microbiome and host gene expression: A review. *Human Genetics*, *140*(5), 747–760. https://doi.org/10.1007/s00439-020-02237-0

CHAPTER 7: LEAKY GUT = LEAKY HORMONES

1. Mudd, A. T., Berding, K., Wang, M., Donovan, S. M., & Dilger, R. N. (2017). Serum cortisol mediates the relationship between fecal *Ruminococcus* and brain N-acetylaspartate in the young pig. *Gut Microbes*, *8*(6), 589–600. https://doi.org/10.1080/19490976.2017.1353849
2. Almand, A. T., Anderson, A. P., Hitt, B. D., Sitko, J. C., Joy, R. M., Easter, B. D., & Almand, E. A. (2022). The influence of perceived stress on the human microbiome. *BMC Research Notes*, *15*(1), article 193. https://doi.org/10.1186/s13104-022-06066-4
3. Madison, A., & Kiecolt-Glaser, J. K. (2019). Stress, depression, diet, and the gut microbiota: human-bacteria interactions at the core of psychoneuroimmunology and nutrition. *Current Opinion in Behavioral Science*, *28*, 105–110. https://doi.org/10.1016/j.cobeha.2019.01.011
4. Org, E., Mehrabian, M., Parks, B. W., Shipkova, P., Liu, X., Drake, T. A., & Lusis, A. J. (2016). Sex differences and hormonal effects on gut microbiota composition in mice. *Gut Microbes*, *7*(4), 313–322. https://doi.org/10.1080/19490976.2016.1203502
5. Nuriel-Ohayon, M., Belogovski, A., Komissarov, S., Ben Izhak, M., Shtossel, O., Neuman, H., Ziv, O., Turjeman, S., Bel, S., Louzoun, Y., & Koren, O. (2021, October 6). *Progesterone supplementation in mice leads to microbiome alterations and weight gain in a sex-specific manner.* (Preprint.) bioRxiv, article 463337. https://doi.org/10.1101/2021.10.06.463337
6. Harada, N. (2018). Role of androgens in energy metabolism affecting on body composition, metabolic syndrome, type 2 diabetes, cardiovascular disease, and longevity: Lessons from a meta-analysis and rodent studies. *Bioscience, Biotechnology, Biochemistry*, *82*(10), 1667–1682. https://doi.org/10.1080/09168451.2018.1490172
7. Collden, H., Landin, A., Wallenius, V., Elebring, E., Fandriks, L., Nilsson, M. E., Ryberg, H., Poutanen, M., Sjögren, K., Vandenput, L., & Ohlsson, C. (2019). The gut microbiota is a major regulator of androgen metabolism in intestinal contents. *American Journal of Physiology: Endocrinology and Metabolism*, *317*(6), E1182–E1192. https://doi.org/10.1152/ajpendo.00338.2019
8. Cross, T.-W. L., Kasahara, K., & Rey, F. E. (2018). Sexual dimorphism of cardiometabolic dysfunction: Gut microbiome in the play? *Molecular Metabolism*, *15*, 70–81.
9. Durmaz, E., Ozmert, E. N., Erkekoglu, P., Giray, B., Derman, O., Hincal, F., & Yurdakök, K. (2010). Plasma phthalate levels in pubertal gynecomastia. *Pediatrics*, *125*(1), e122–e129. https://doi.org/10.1542/peds.2009-0724

10. Koren, O., Goodrich, J. K., Cullender, T. C., Spor, A., Laitinen, K., Bäckhed, H. K., Gonzalez, A., Werner, J. J., Angenent, L. T., Knight, R., Bäckhed, F., Isolauri, E., Salminen, S., & Ley, R. E. (2012). Host remodeling of the gut microbiome and metabolic changes during pregnancy. *Cell*, *150*(3), 470–480.
11. Mueller, S., Saunier, K., Hanisch, C., Norin, E., Alm, L., Midtvedt, T., Cresci, A., Silvi, S., Orpianesi, C., Verdenelli, M. C., Clavel, T., Koebnick, C., Zunft, H.-J. F., Doré, J., & Blaut, M. (2006). Differences in fecal microbiota in different European study populations in relation to age, gender, and country: A cross-sectional study. *Applied and Environmental Microbiology*, *72*(2), 1027–1033. https://doi.org/10.1128/AEM.72.2.1027-1033.2006
12. Peters, B. A., Lin, J., Qi, Q., Usyk, M., Isasi, C. R., Mossavar-Rahmani, Y., Derby, C. A., Santoro, N., Perreira, K. M., Daviglus, M. L., Kominiarek, M. A., Cai, J., Knight, R., Burk, R. D., & Kaplan, R. C. (2022). Menopause is associated with an altered gut microbiome and estrobolome, with implications for adverse cardiometabolic risk in the Hispanic community health study/study of Latinos. *mSystems*, 7(3), e0027322. https://doi.org/10.1128/msystems.00273-22
13. Kaliannan, K., Robertson, R. C., Murphy, K., Stanton, C., Kang, C., Wang, B., Hao, L., Bhan, A. K., & Kang, J. X. (2018). Estrogen-mediated gut microbiome alterations influence sexual dimorphism in metabolic syndrome in mice. *Microbiome*, *6*(1), 205.
14. Homma, H., Hoy, E., Xu, D.-Z., Lu, Q., Feinman, R., & Deitch, E. A. (2005). The female intestine is more resistant than the male intestine to gut injury and inflammation when subjected to conditions associated with shock states. *American Journal of Physiology: Gastrointestinal and Liver Physiology*, *288*(3), G466–G472. https://doi.org/10.1152/ajpgi.00036.2004
15. Shieh, A., Epeldegui, M., Karlamangla, A. S., & Greendale, G. A. (2020). Gut permeability, inflammation, and bone density across the menopause transition. *JCI Insight*, *5*(2), e134092. https://doi.org/10.1172/jci.insight.134092
16. Baker, J. M., Al-Nakkash, L., & Herbst-Kralovetz, M. M. (2017). Estrogen-gut microbiome axis: Physiological and clinical implications. *Maturitas*, *103*, 45–53. https://doi.org/10.1016/j.maturitas.2017.06.025
17. Adlercreutz, H., Pulkkinen, M. O., Hämäläinen, E. K., & Korpela, J. T. (1984). Studies on the role of intestinal bacteria in metabolism of synthetic and natural steroid hormones. *Journal of Steroid Biochemistry*, *20*(1), 217–229. https://doi.org/10.1016/0022-4731(84)90208-5
18. Plottel, C. S., & Blaser, M. J. (2011). Microbiome and malignancy. *Cell Host & Microbe*. *10*(4), 324–335. https://doi.org/10.1016/j.chom.2011.10.003
19. Ervin, S. M., Li, H., Lim, L., Roberts, L. R., Liang, X., Mani, S., & Redinbo, M. R. (2019). Gut microbial β-glucuronidases reactivate estrogens as components of the estrobolome that reactivate estrogens. *Journal of Biological Chemistry*, *294*(49), 18586–18599. https://doi.org/10.1074/jbc.RA119.010950
20. Alizadehmohajer, N., Shojaeifar, S., Nedaeinia, R., Esparvarinha, M., Mohammadi, F., Ferns, G. A., Ghayour-Mobarhan, M., Manian, M., & Balouchi, A. (2020). Association between the microbiota and women's cancers—Cause or consequences? *Biomedicine & Pharmacotherapy*, *127*, 110203.

21. Baker, J. M., Al-Nakkash, L., & Herbst-Kralovetz, M. M. (2017). Estrogen-gut microbiome axis: Physiological and clinical implications. *Maturitas*, *103*, 45–53. https://doi.org/10.1016/j.maturitas.2017.06.025
22. Baker, J. M., Al-Nakkash, L., & Herbst-Kralovetz, M. M. (2017). Estrogen-gut microbiome axis: Physiological and clinical implications. *Maturitas*, *103*, 45–53. https://doi.org/10.1016/j.maturitas.2017.06.025
23. Anderson, G. (2019). Endometriosis pathoetiology and pathophysiology: Roles of vitamin A, estrogen, immunity, adipocytes, gut microbiome and melatonergic pathway on mitochondria regulation. *Biomolecular Concepts*, *10*(1), 133–149. https://doi.org/10.1515/bmc-2019-0017
24. Ata, B., Yildiz, S., Turkgeldi, E., Brocal, V. P., Dinleyici, E. C., Moya, A., & Urman, B. (2019). The Endobiota Study: Comparison of vaginal, cervical and gut microbiota between women with stage 3/4 endometriosis and healthy controls. *Scientific Reports*, *9*(1), 2204.
25. Rosean, C. B., Bostic, R. R., Ferey, J.C.M., Feng, T.-Y., Azar, F. N., Tung, K. S., Dozmorov, M. G., Smirnova, E., Bos, P. D., & Rutkowski, M. R. (2019). Pre-existing commensal dysbiosis is a host-intrinsic regulator of tissue inflammation and tumor cell dissemination in hormone receptor-positive breast cancer. *Cancer Research*, *79*(14), 3662–3675. https://doi.org/10.1158/0008-5472.CAN-18-3464
26. Parida, S., & Sharma, D. (2020). Microbial alterations and risk factors of breast cancer: Connections and mechanistic insights. *Cells*, *9*(5), 1091.
27. Mikó, E., Kovács, T., Sebő, É., Tóth, J., Csonka, T., Ujlaki, G., Sipos, A., Szabó, J., Méhes, G., & Bai, P. (2019). Microbiome–microbial metabolome–cancer cell interactions in breast cancer—familiar, but unexplored. *Cells*, *8*(4), 293. https://doi.org/10.3390/cells8040293
28. Kwa, M., Plottel, C. S., Blaser, M. J., & Adams, S. (2016). The intestinal microbiome and estrogen receptor–positive female breast cancer. *Journal of the National Cancer Institute*, *108*(8), article djw029. https://doi.org/10.1093/jnci/djw029
29. Endocrine Society. (2022, June 11). *Probiotic bacteria may enhance tamoxifen effectiveness in treatment of ER+ breast cancer.* https://admin.endocrine.org/news-and-advocacy/news-room/2022/probiotic-bacteria-may-enhance-tamoxifen-effectiveness-in-treatment-of-er-breast-cancer
30. Parida, S., & Sharma, D. (2019). The microbiome-estrogen connection and breast cancer risk. *Cells*, *8*(12), 1642. https://doi.org/10.3390/cells8121642
31. Toumazi, D., El Daccache, S., & Constantinou, C. (2021). An unexpected link: The role of mammary and gut microbiota on breast cancer development and management (review). *Oncology Reports*, *45*(5), 80. https://doi.org/10.3892/or.2021.8031
32. Tzeng, A., Sangwan, N., Jia, M., Liu, C.-C., Keslar, K. S., Downs-Kelly, E., Fairchild, R. L., Al-Hilli, Z., Grobmyer, S. R., & Eng, C. (2021). Human breast microbiome correlates with prognostic features and immunological signatures in breast cancer. *Genome Medicine*, *13*(1), 60. https://doi.org/10.1186/s13073-021-00874-2
33. Mikó, E., Kovács, T., Sebő, É., Tóth, J., Csonka, T., Ujlaki, G., Sipos, A., Szabó,

J., Méhes, G., & Bai, P. (2019). Microbiome–microbial metabolome–cancer cell interactions in breast cancer—familiar, but unexplored. *Cells*, *8*(4), 293. https://doi.org/10.3390/cells8040293

34. Vital, M., Howe, A. C., & Tiedje, J. M. (2014). Revealing the bacterial butyrate synthesis pathways by analyzing (meta) genomic data. *mBio*, *5*(2), e00889–14.
35. Shrode, R. L., Knobbe, J. E., Cady, N., Yadav, M., Hoang, J., Cherwin, C., Curry, M., Garje, R., Vikas, P., Sugg, S., Phadke, S., Filardo, E., & Mangalam, A. K. (2023). Breast cancer patients from the Midwest region of the United States have reduced levels of short-chain fatty acid–producing gut bacteria. *Scientific Reports*, *13*(1), 526. https://doi.org/10.1038/s41598-023-27436-3
36. Wang, Q., Zhao, L., Han, L., Fu, G., Tuo, X., Ma, S., Li, Q., Wang, Y., Liang, D., Tang, M., Sun, C., Wang, Q., Song, Q., & Li, Q. (2020). The differential distribution of bacteria between cancerous and noncancerous ovarian tissues in situ. *Journal of Ovarian Research*, *13*(1), 8.
37. Park, G. B., Chung, Y. H., & Kim, D. (2017). Induction of galectin-1 by TLR-dependent PI3K activation enhances epithelial-mesenchymal transition of metastatic ovarian cancer cells. *Oncology Reports*, *37*(5), 3137–3145.
38. Kashani, B., Zandi, Z., Bashash, D., Zaghal, A., Momeny, M., Poursani, E. M., Pourbagheri-Sigaroodi, A., Mousavi, S. A., & Ghaffari, S. H. (2020). Small molecule inhibitor of TLR4 inhibits ovarian cancer cell proliferation: new insight into the anticancer effect of TAK-242 (resatorvid). *Cancer Chemotherapy and Pharmacology*, *85*(1), 47–59.
39. Łaniewski, P., Ilhan, Z. E., & Herbst-Kralovetz, M. M. (2020). The microbiome and gynaecological cancer development, prevention and therapy. *Nature Reviews Urology*, *17*(4), 232–250.
40. Dhingra, A., Sharma, D., Kumar, A., Singh, S., & Kumar, P. (2022). Microbiome and development of ovarian cancer. *Endocrine, Metabolic & Immune Disorders Drug Targets*, *22*(11), 1073–1090. https://doi.org/10.2174/1871530322666220509034847
41. Terao, Y., Nishida, J., Horiuchi, S., Rong, F., Ueoka, Y., Matuda, T., Kato, H., Furugen, Y., Yoshida, K., Kato, K., & Wake, N. (2001). Sodium butyrate induces growth arrest and senescence-like phenotypes in gynecologic cancer cells. *International Journal of Cancer*, *94*(2), 257–267.
42. Lamb, R., Ozsvari, B., Lisanti, C. L., Tanowitz, H. B, Howell, A., Martinez-Outschoom, U. E., Sotgia, F., & Lisanti, M. P. (2015). Antibiotics that target mitochondria effectively eradicate cancer stem cells, across multiple tumor types: Treating cancer like an infectious disease. *Oncotarget*, *6*(7), 4569–4584.
43. Wang, W., Qin, X., Hu, D., Huang, J., Guo, E., Xiao, R., Li, W., & Sun, C. (2022). *Akkermansia* supplementation reverses the tumor-promoting effect of the fecal microbiota transplantation in ovarian cancer. *Cell Reports*, *41*(13), article 111890. https://doi.org/10.1016/j.celrep.2022.111890
44. Montjean, D., Neyroud, A. S., Yefimova, M. G., Benkhalifa, M., Cabry, R., & Ravel, C. (2022). Impact of endocrine disruptors upon non-genetic inheritance. *International Journal of Molecular Sciences*, *23*(6), 3350. https://doi.org/10.3390/ijms23063350

45. Eskenazi, B., Ames, J., Rauch, S., Signorini, S., Brambilla, P., Mocarelli, P., Siracusa, C., Holland, N., & Warner, M. (2021). Dioxin exposure associated with fecundability and infertility in mothers and daughters of Seveso, Italy. *Human Reproduction*, *36*(3), 794–807. https://doi.org/10.1093/humrep/deaa324
46. Kirchhof, M. G., & de Gannes, G. C. (2013). The health controversies of parabens. *Skin Therapy Letter*, *18*(2), 5–7.
47. Golden, R., Gandy, J., & Vollmer, G. (2005). A review of the endocrine activity of parabens and implications for potential risks to human health. *Critical Reviews in Toxicology*, *35*(5), 435–458. https://doi.org/10.1080/10408440490920104
48. Forte, M., Di Lorenzo, M., Carrizzo, A., Valiante, S., Vecchione, C., Laforgia, V., & De Falco, M. (2016). Nonylphenol effects on human prostate non tumorigenic cells. *Toxicology*, *357–358*, 21–32. https://doi.org/10.1016/j.tox.2016.05.024
49. Alwadi, D., Felty, Q., Roy, D., Yoo, C., & Deoraj, A. (2022). Environmental phenol and paraben exposure risks and their potential influence on the gene expression involved in the prognosis of prostate cancer. *International Journal of Molecular Sciences*. *23*(7), 3679. https://doi.org/10.3390/ijms23073679
50. López-Carrillo, L., Hernández-Ramírez, R. U., Calafat, A. M., Torres-Sánchez, L., Galván-Portillo, M., Needham, L. L., Ruiz-Ramos, R., & Cebrián, M. E. (2010). Exposure to phthalates and breast cancer risk in northern Mexico. *Environmental Health Perspectives*, *118*(4), 539–544. https://doi.org/10.1289/ehp.0901091
51. Ahern, T. P., Broe, A., Lash, T. L., Cronin-Fenton, D. P., Ulrichsen, S. P., Christiansen, P. M., Cole, B. F., Tamimi, R. M., Sørensen, H. T., & Damkier, P. (2019). Phthalate exposure and breast cancer incidence: A Danish nationwide cohort study. *Journal of Clinical Oncology*, *37*(21), 1800–1809. https://doi.org/10.1200/JCO.18.02202
52. Radke, E. G., Braun, J. M., Meeker, J. D., & Cooper, G. S. (2018). Phthalate exposure and male reproductive outcomes: A systematic review of the human epidemiological evidence, *Environment International*, *121*(1), 764–793. https://doi.org/10.1016/j.envint.2018.07.029
53. Peng, M. Q., Karvonen-Gutierrez, C. A., Herman, W. H., Mukherjee, B., & Park, S. K. (2023). Phthalates and incident diabetes in midlife women: The Study of Women's Health Across the Nation (SWAN). *The Journal of Clinical Endocrinology & Metabolism*, dgad033. (Preprint ahead of publication.) https://doi.org/10.1210/clinem/dgad033
54. Edwards, L., McCray, N. L., VanNoy, B. N., Yau, A., Geller, R. J., Adamkiewicz, G., & Zota, A. R. (2022). Phthalate and novel plasticizer concentrations in food items from U.S. fast food chains: A preliminary analysis. *Journal of Exposure Science & Environmental Epidemiology*, *32*(1), 366–373. https://doi.org/10.1038/s41370-021-00392-8
55. Gan, W., Zhou, M., Xiang, Z., Han, X., & Li, D. (2015). Combined effects of nonylphenol and bisphenol A on the human prostate epithelial cell line RWPE-1. *International Journal of Environmental Research and Public Health*, *12*(4), 4141–4155. https://doi.org/10.3390/ijerph120404141
56. Crobeddu, B., Ferraris, E., Kolasa, E., & Plante, I. (2019). Di(2-ethylhexyl)

phthalate (DEHP) increases proliferation of epithelial breast cancer cells through progesterone receptor dysregulation. *Environmental Research, 173*, 165–173. https://doi.org/10.1016/j.envres.2019.03.037
57. Wetherill, Y. B., Akingbemi, B. T., Kanno, J., McLachlan, J. A., Nadal, A., Sonnenschein, C., Watson, C. S., Zoeller, R. T., & Belcher, S. M. (2007). *In vitro* molecular mechanisms of bisphenol A action. *Reproductive Toxicology, 24*(2), 178–198. https://doi.org/10.1016/j.reprotox.2007.05.010
58. Lee, H. J., Chattopadhyay, S., Gong, E.-Y., Ahn, R. S., & Lee, K. (2003). Antiandrogenic effects of bisphenol A and nonylphenol on the tunction of androgen receptor. *Toxicological Sciences, 75*(1), 40–46. https://doi.org/10.1093/toxsci/kfg150
59. Seachrist, D. D., Bonk, K. W., Ho, S.-M., Prins, G. S., Soto, A. M., & Keri, R. A. (2016). A review of the carcinogenic potential of bisphenol A. *Reproductive Toxicology, 59*, 167–182. https://doi.org/10.1016/j.reprotox.2015.09.006
60. Nicolopoulou-Stamati, P., Maipas, S., Kotampasi, C., Stamatis, P., & Hens, L. (2016). Chemical pesticides and human health: The urgent need for a new concept in agriculture. *Frontiers in Public Health, 4*, 148. https://doi.org/10.3389/fpubh.2016.00148
61. Liu, J., Zhao, M., Zhuang, S., Yang, Y., Yang, Y., & Liu, W. (2012). Low concentrations of *o,p*⊠-DDT inhibit gene expression and prostaglandin synthesis by estrogen receptor–independent mechanism in rat ovarian cells. *PLOS ONE*, 7(11), e49916. https://doi.org/10.1371/journal.pone.0049916
62. Cohn, B. A., Cirillo, P. M., & Terry, M. B. (2019). DDT and breast cancer: Prospective study of induction time and susceptibility windows. *Journal of the National Cancer Institute, 111*(8), 803–810. https://doi.org/10.1093/jnci/djy198
63. Perry, M. J., Young, H. A., Grandjean, P., Halling, J., Petersen, M. S., Martenies, S. E., Karimi, P., & Weihe, P. (2016). Sperm aneuploidy in Faroese men with lifetime exposure to dichlorodiphenyldichloroethylene (*p,p*⊠-DDE) and polychlorinated biphenyl (PCB) pollutants. *Environmental Health Perspectives, 124*(7), 951–956.
64. Maness, S. C., McDonnell, D. P., & Gaido, K. W. (1998). Inhibition of androgen receptor–dependent transcriptional activity by DDT isomers and methoxychlor in HepG2 human heptaoma cells. *Toxicology and Applied Pharmacology, 151*(1),135–142. https://doi.org/10.1006/taap.1998.8431
65. Maness, S. C., McDonnell, D. P., & Gaido, K. W. (1998). Inhibition of androgen receptor–dependent transcriptional activity by DDT isomers and methoxychlor in HepG2 human heptaoma cells. *Toxicology and Applied Pharmacology, 151*(1),135–142. https://doi.org/10.1006/taap.1998.8431
66. Xiagedeer, B., Hou, X., Zhang, Q., Hu, H., Kang, C., Xiao, Q., & Hao, W. (2020). Maternal chlormequat chloride exposure disrupts embryonic growth and produces postnatal adverse effects. *Toxicology, 442*, article 152534.
67. Evans, S., Temkin, A., & Naidenko, O. (2023, January 31). *EWG investigation: Dangerous agricultural chemical chlormequat found in popular oat-based products.* Environmental Working Group. https://www.ewg.org/research/ewg-investigation-dangerous-agricultural-chemical-chlormequat-found-popular-oat-based

CHAPTER 8: CIGARETTES, MEAT, AND CHEESE: THE SECRETS OF LONGEVITY ARE NOT WHAT YOU THINK

1. Rampelli, S., Soverini, M., D'Amico, F., Barone, M., Tavella, T., Monti, D., Capri, M., Astolfi, A., Brigidi, P., Biagi, E., Franceschi, C., Turroni, S., & Candela, M. (2020). Shotgun metagenomics of gut microbiota in humans with up to extreme longevity and the increasing role of xenobiotic degradation. *mSystems*, *5*(2), e00124-20. https://doi.org/10.1128/mSystems.00124-20
2. Newman, S. J. (2018). Plane inclinations: A critique of hypothesis and model choice in Barbi et al. *PLOS Biology*, *16*(12), e3000048. https://doi.org/10.1371/journal.pbio.3000048
3. Newman, S. J. (2020). Supercentenarian and remarkable age records exhibit patterns indicative of clerical errors and pension fraud. bioRxiv, article 704080. (Preprint ahead of publication.)
4. CIA. *CIA World Factbook*, 2013. https://www.cia.gov/the-world-factbook/
5. Chetty, R., Stepner, M., Abraham, S., Lin, S., Scuderi, B., Turner, N., Bergeron, A., & Cutler, D. (2016). The association between income and life expectancy in the United States, 2001–2014. *The Journal of the American Medical Association*, *315*(16), 1750–1766. https://doi.org/10.1001/jama.2016.4226
6. Poulain, M., Herm, A., & Pes, G. (2013). The Blue Zones: Areas of exceptional longevity around the world. *Vienna Yearbook of Population Research*, *11*, 87–108. https://doi.org/10.1553/populationyearbook2013s87
7. Panagiotakos, D. B., Chrysohoou, C., Siasos, G., Zisimos, K., Skoumas, J., Pitsavos, C., & Stefanadis, C. (2011). Sociodemographic and lifestyle statistics of oldest old people (>80 years) living in Ikaria island: The Ikaria study. *Cardiology Research and Practice*, article 679187. https://doi.org/10.4061/2011/679187
8. Martínez-González, M. A., García-López, M., Bes-Rastrollo, M., Toledo, E., Martínez-Lapiscina, E. H., Delgado-Rodriguez, M., Vazquez, Z., Benito, S., & Beunza, J. J. (2011). Mediterranean diet and the incidence of cardiovascular disease: A Spanish cohort. *Nutrition, Metabolism & Cardiovascular Diseases*, *21*(4), 237–244. https://www.ncbi.nlm.nih.gov/pubmed/20096543
9. Schünke, M., Schumacher, U., & Tillmann, B. (1985). Lectin-Binding in Normal and Fibrillated Articular Cartilage of Human Patellae. *Virchows Archiv A, Pathological Anatomy and Histophatology*, *407*(2), 221–231. https://www.ncbi.nlm.nih.gov/m/pubmed/3927585/?i=5&from=/23214295/related
10. Sardu, C., Cocco, E., Mereu, A., Massa, R., Cuccu, A., Marrosu, G., & Contu, P. (2012). Population based study of 12 automimmune diseases in Sardinia, Italy: Prevalence and comorbidity. *PLOS ONE*, *7*(3), e32487. https://journals.plos.org/plosone/article?id=10.1371/journal.pone.0032487
11. Vasto, S., Scapagnini, G., Rizzo, C., Monastero, R., Marchese, A., & Caruso, C. (2012). Mediterranean diet and longevity in Sicily: Survey in a Sicani Mountains population. *Rejuvenation Research*, *15*(2), 184–188. https://doi.org/10.1089/rej.2011.1280
12. Demmer, E., Van Loon, M. D., Rivera, N., Rogers, T. S., Gertz, E. R., German, J. B., Smilowitz, J. T., & Zivkovic, A. M. (2016). Addition of a dairy fraction rich in milk fat globule membrane to a high–saturated fat meal reduces the

postprandial insulinaemic and inflammatory response in overweight and obese adults. *Journal of Nutritional Science*, *5*, e14. https://www.ncbi.nlm.nih.gov/pmc/articles/PMC4791522/

13. Ji, X., Xu, W., Cui, J., Ma, Y., & Zhou, S. (2019). Goat and buffalo milk fat globule membranes exhibit better effects at inducing apoptosis and reduction the viability of HT-29 cells. *Scientific Reports*, *9*(1), article 2577. https://www.nature.com/articles/s41598-019-39546-y
14. Ardisson Korat, A. V. (2018). *Dairy products and cardiometabolic health outcomes* (Publication No. 28225720) [Doctoral dissertation, Harvard University]. ProQuest.
15. Pirinen, E., Kuulasmaa, T., Pietilä, M., Heikkinen, S., Tusa, M., Itkonen, P., Boman, S., Skommer, J., Virkamäki, A., Hohtola, E., Kettunen, M., Fatrai, S., Kansanen, E., Koota, S., Niiranen, K., Parkkinen, J., Levonen, A.-L., Ylä-Herttuala, S., Hiltunen, J. K., . . . Laakso, M. (2020). Enhanced polyamine catabolism alters homeostatic control of white adipose tissue mass, energy expenditure, and glucose metabolism. *Molecular and Cellular Biology*, *27*(13), 4953–4967. https://mcb.asm.org/content/27/13/4953
16. Koskinen, T. T., Virtanen, H.E.K., Voutilainen, S., Tuomainen, T. P., Mursu, J., & Virtanen, J. K. (2018). Intake of fermented and non-fermented dairy products and risk of incident CHD: The Kuopio Ischaemic Heart Disease Risk Factor Study. *British Journal of Nutrition*, *120*(11), 1288–1297. https://doi.org/10.1017/S0007114518002830
17. Tognon, G., Nilsson, L. M., Shungin, D., Lissner, L., Jansson, J.-H., Renström, F., Wennberg, M., Winkvist, A., & Johansson, I. (2017). Nonfermented milk and other dairy products: Associations with all-cause mortality. *The American Journal of Clinical Nutrition*, *105*(6), 1502–1511. https://doi.org/10.3945/ajcn.116.140798
18. Lallès, J. P. (2016). Dairy products and the French paradox: Could alkaline phosphatases play a role? *Medical Hypotheses*, *92*, 7–11. https://doi.org/10.1016/j.mehy.2016.04.033
19. Petyaev, I. M., & Bashmakov, Y. K. (2012). Could cheese be the missing piece in the French paradox puzzle? *Medical Hypotheses*, *79*(6), 746–749. https://doi.org/10.1016/j.mehy.2012.08.018
20. Hallajzadeh, J., Eslami, R. D., & Tanomand, A. (2021). Effect of *Lactobacillus delbrueckii* subsp. *lactis* PTCC1057 on serum glucose, fetuin-A, and sestrin 3 levels in streptozotocin-induced diabetic mice. *Probiotics and Antimicrobial Proteins*, *13*(2), 383–389. https://doi.org/10.1007/s12602-020-09693-0
21. Anggraini, H., Tongkhao, K., & Chanput, W. (2021). Reducing milk allergenicity of cow, buffalo, and goat milk using lactic acid bacteria fermentation. *AIP Conference Proceedings*, article 010001.
22. Nieddu, A., Vindas, L., Errigo, A., Vindas, J., Pes, G. M., & Dore, M. P. (2020). Dietary habits, anthropometric features and daily performance in two independent long-lived populations from *Nicoya peninsula* (Costa Rica) and *Ogliastra* (Sardinia). *Nutrients*, *12*(6), 1621. https://doi.org/10.3390/nu12061621
23. Lutsiv, T., McGinley, J. N., Neil-McDonald, E. S., Weir, T. L., Foster, M. T., & Thompson, H. J. (2022). Relandscaping the gut microbiota with a whole food:

Dose-response effects to common bean. *Foods*, *11*(8), 1153. https://doi.org/10.3390/foods11081153

24. Garcia-Mantrana, I., Selma-Royo, M., Alcantara, C., & Collado, M. C. (2018). Shifts on gut microbiota associated to Mediterranean diet adherence and specific dietary intakes on general adult population. *Frontiers in Microbiology*, *9*, article 890. https://doi.org/10.3389/fmicb.2018.00890
25. Robine, J. M., Herrmann, F. R., Arai, Y., Willcox, D. C., Gondo, Y., Hirose, N., Suzuki, M., & Saito, Y. (2012). Exploring the impact of climate on human longevity. *Experimental Gerontology*, *47*(9), 660–671. https://doi.org/10.1016/j.exger.2012.05.009
26. Willcox, B., Willcox, D. C., & Suzuki, M. (2004). *The Okinawa diet plan.* Three Rivers Press.
27. Korpela, K., Flint, H. J., Jonstone, A. M., Lappi, J., Poutanen, K., Dewulf, E., Delzenne, N., de Vos, W. M., & Salonen, A. (2014). Gut microbiota signatures predict host and microbiota responses to dietary interventions in obese individuals. *PLOS ONE*, *9*(6), e90702. http://www.oalib.com/references/8108647
28. Modinham, C. L., Frost, G. S., & Robertson, M. D. (2010). Acute ingestion of resistant starch reduces food intake in healthy adults. *British Journal of Nutrition*, *103*(6), 917–922. http://journals.cambridge.org/action/displayAbstract?fromPage=online&aid=7358712&fileId=S0007114509992534
29. Nilsson, A. C., Ostman, E. M., Holst, J. J., & Björck, I.M.E. (2008). Including indigestible carbohydrates in the evening meal of healthy subjects improves glucose tolerance, lowers inflammatory markers, and increases satiety after a subsequent standardized breakfast. *The Journal of Nutrition*, *138*(4), 732–739. http://www.ncbi.nlm.nih.gov/pubmed/18356328
30. Hou, W.-C., Chen, Y.-C., Chen, H.-J., Lin, Y.-H., Yang, L.-L., & Lee, M.-H. (2001). Antioxidant activities of trypsin inhibitor, a 33 Kda root storage protein of sweet potato (*Ipomoea batatas* (L.) Lam cv. Tainong 57). *Journal of Agricultural and Food Chemistry*, *49*(6), 2978–2981.
31. Dini, I., Tenore, G. C., & Dini, A. (2006). New polyphenol derivative in *Ipomoea batatas* tubers and its antioxidant activity. *Journal of Agricultural and Food Chemistry*, *54*(23), 8733–8737.
32. Kano, M., Takayanagi, T., Harada, K., Makino, K., & Ishikawa, F. (2005). Antioxidative activity of anthocyanins from purple sweet potato, *Ipomoea batatas* cultivar Ayamurasaki. *Bioscience, Biotechnology, Biochemistry*, *69*(5), 979–988.
33. Kurata, R., Adachi, M., Yamakawa, O., & Yoshimoto, M. (2007). Growth suppression of human cancer cells by polymorphenolics from sweet potato (*Ipomoea batatas* L.) leaves. *Journal of Agricultural and Food Chemistry*, *55*(1), 185–190.
34. Frolinger, T., Sims, S., Smith, C., Wang, J., Cheng, H., Faith, J., Ho, L., Hao, K., & Pasinetti, G. M. (2019). The gut microbiota composition affects dietary polyphenols–mediated cognitive resilience in mice by modulating the bioavailability of phenolic acids. *Scientific Reports*, *3*(Suppl. 1), article 3546. https://doi.org/10.1038/s41598-019-39994-6
35. Ribeiro Pereira, P., Bertozzi de Aquino Mattos, É., Nitzsche Teixeira Fernandes Corrêa, A. C., Vericimo M. A., & Flosi Paschoalin, V. M. (2020). Anticancer and immunomodulatory benefits of taro (*Colocasia esculenta*) corms, an

underexploited tuber crop. *International Journal of Molecular Sciences*, *22*(1), 265. https://doi.org/10.3390/ijms22010265

36. Willcox, B., Willcox, D. C., & Suzuki, M. (2004). *The Okinawa diet plan.* Three Rivers Press.
37. Barbieri, F., Tabanelli, G., Montanari, C., Dall'Osso, N., Šimat, V., Možina, S. S., Baños, A., Özogul, F., Bassi, D., Fontana, C., & Gardini, F. (2021). Mediterranean spontaneously fermented sausages: Spotlight on microbiological and quality features to exploit their bacterial biodiversity. *Foods*, *10*(11), 2691. https://doi.org/10.3390/foods10112691
38. Ruiz-Capillas, C., & Jiménez-Colmenero, F. (2004). Biogenic amines in meat and meat products. *Critical Reviews in Food Science and Nutrition*, *44*(7–8), 489–499. https://doi.org/10.1080/10408690490489341
39. Depauw, S., Bosch, G., Hesta, M., Whitehouse-Tedd, K., Hendriks, W. H., Kaandorp, J., & Janssens, G. P. (2012). Fermentation of animal components in strict carnivores: A comparative study with cheetah fecal inoculum. *Journal of Animal Science*, *90*(8), 2540–2548. https://doi.org/10.2527/jas.2011-4377
40. Willcox, D. C., Willcox, B. J., Todoriki, H., & Suzuki, M. (2009). The Okinawan diet: Health implications of a low-calorie, nutrient-dense, antioxidant-rich dietary pattern low in the glycemic load. *Journal of the American College of Clinical Nutrition*, *28*(Suppl.), S500–S516. https://doi.org/10.1080/07315724.2009.10718117
41. Mei, X.-D., Cao, Y.-F., Che, Y.-Y., Li, J., Shang, Z.-P., Zhao, W.-J., Qiao, Y.-J., & Zhang, J.-Y. (2019). Danshen: A phytochemical and pharmacological overview. *Chinese Journal of Natural Medicines*, *17*(1), 59–80. https://doi.org/10.1016/S1875-5364(19)30010-X
42. Matsunami, K., & Otsuka, H. (2018). Okinawan subtropical plants as a promising resource for novel chemical treasury. *Chemical & Pharmaceutical Bulletin*, *66*(5), 519–526. https://doi.org/10.1248/cpb.c17-00831
43. Shinzato, C., Inoue, M., & Kusakabe, M. (2014). A snapshot of a coral "holobiont": A transcriptome assembly of the scleractinian coral, porites, captures a wide variety of genes from both the host and symbiotic zooxanthellae. *PLOS ONE*, *9*(1), e85182. https://doi.org/10.1371/journal.pone.0085182
44. Taguchi, C., Kishimoto, Y., Fukushima, Y., Kondo, K., Yamakawa, M., Wada, K., & Nagata, C. (2020). Dietary intake of total polyphenols and the risk of all-cause and specific-cause mortality in Japanese adults: The Takayama Study. *European Journal of Nutrition*, *59*(3), 1263–1271. https://doi.org/10.1007/s00394-019-02136-9
45. Willcox, D. C., Willcox, B. J., Todoriki, H., & Suzuki, M. (2009). The Okinawan diet: Health implications of a low-calorie, nutrient-dense, antioxidant-rich dietary pattern low in glycemic load. *Journal of the American College of Nutrition*, *28*(Suppl.), S500–S516. https://doi.org/10.1080/07315724.2009.10718117
46. Biasi, F., Guina, T., Maina, M., Cabboi, B., Deiana, M., Tuberoso, C. I., Calfapietra, S., Chiarpotto, E., Sottero, B., Gamba, P., Gargiulo, S., Brunetto, V., Testa, G., Dessì, M. A., Poli, G., & Leonarduzzi, G. (2013). Phenolic compounds present in Sardinian wine extracts protect against the production of

inflammatory cytokines induced by oxysterols in CaCo-2 human enterocyte-like cells. *Biochemical Pharmacology*, *86*(1), 138–145. https://doi.org/10.1016/j.bcp.2013.03.024

47. Nieddu, A., Vindas, L., Errigo, A., Vindas, J., Pes, G. M., & Dore, M. P. (2020). Dietary habits, anthropometric features and daily performance in two independent long-lived populations from *Nicoya peninsula* (Costa Rica) and *Ogliastra* (Sardinia). *Nutrients*, *12*(6), 1621. https://doi.org/10.3390/nu12061621
48. Panagiotakos, D. B., Chrysohoou, C., Siasos, G., Zisimos, K., Skoumas, J., Pitsavos, C., & Stefanadis, C. (2011). Sociodemographic and lifestyle statistics of oldest old people (>80 years) living in Ikaria island: The Ikaria study. *Cardiology Research and Practice*, article 679187. https://doi.org/10.4061/2011/679187
49. de Lorgeril, M., Salen, P., Martin, J. L., Monjaud, I., Delaya, J., & Mamelle, N. (1999). Mediterranean diet, traditional risk factors, and the rate of cardiovascular complications after myocardial infarction: Final report of the Lyon Diet Heart Study. *Circulation*, *99*(6), 779–785. https://doi.org/10.1161/01.CIR.99.6.779
50. Chen, W., Yu, Y., Liu, Y., Song, C., Chen, H., Tang, C., Song, Y., & Zhang, X. (2022). Ursolic acid regulates gut microbiota and corrects the imbalance of Th17/Treg cells in T1DM rats. *PLOS ONE*, *17*(11), e0277061. https://doi.org/10.1371/journal.pone.0277061
51. Mark, K. A., Dumas, K. J., Bhaumik, D., Schilling, B., Davis, S., Oron, T. R., Sorenen, D. J., Lucanic, M., Brem, R. B., Melov, S., Ramanathan, A., Gibson, B. W., & Lithgow, G. J. (2016). Vitamin D promotes protein homeostasis and longevity via the stress response pathway genes *skn-1*, *ire-1*, and *xbp-1*. *Cell Reports*, *17*(5), 1227–1237. https://doi.org/10.1016/j.celrep.2016.09.086
52. Thomas, R. L., Jiang, L., Adams, J. S., Xu, Z. Z., Shen, J., Janssen, S., Ackermann, G., Vanderschueren, D., Pauwels, S., Knight, R, Orwoll, E. S., & Kado, D. M. (2020). Vitamin D metabolites and the gut microbiome in older men. *Nature Communications*, *11*(2), 5997. https://doi.org/10.1038/s41467-020-19793-8
53. Singh, P., Rawat, A., Alwakeel, M., Sharif, E., & Al Khodor, S. (2020). The potential role of vitamin D supplementation as a gut microbiota modifier in healthy individuals. *Scientific Reports*, *10*(1), article 21641. https://doi.org/10.1038/s41598-020-77806-4
54. Kanasuo, E., Siiskonen, H., Haimakainen, S., Komulainen, J., & Harvima, I. T. (2023). Regular use of vitamin D supplement is associated with fewer melanoma cases compared to non-use: A cross-sectional study in 498 adult subjects at risk of skin cancers. *Melanoma Research*, *33*(2), 126–135. https://doi.org/10.1097/CMR.0000000000000870
55. Shepherds Purse. (2020, July 16). *Six fascinating facts about sheep milk.* https://blog.shepherdspurse.co.uk/blog/six-facts-benefits-sheep-milk
56. Yoshida, Y., Sakane, N., Umekawa, T., Kogure, A., Kondo, M., Kumamoto, K., Kawada, T., Nagase, I., & Saito, M. (1999). Nicotine induces uncoupling protein 1 in white adipose tissue of obese mice. *International Journal of Obesity and Related Metabolic Disorders*, *23*(6), 570–575. https://pubmed.ncbi.nlm.nih.gov/10411229/

57. Mappin-Kasirer, B., Pan, H., Lewington, S., Kizza, J., Gray, R., Clarke, R., & Peto, R. (2020). Tobacco smoking and the risk of Parkinson disease: A 65-year follow-up of 30,000 Male British doctors. *Neurology*, *94*(20), e2132–e2138. https://n.neurology.org/content/94/20/e2132
58. van Duijn, C. M., & Hofman, A. (1991). Relation between nicotine intake and Alzheimer's disease. *The British Medical Journal*, *302*(6791), 1491–1494. https://www.ncbi.nlm.nih.gov/pmc/articles/PMC1670208/
59. Sardi, B. (2001). The two faces of vitamin C. *Science*, *293*(5537), 1993–1995. https://doi.org/10.1126/science.293.5537.1993. Erratum (2001) in: *Science*, *294*(5543), 788.
60. Willcox, B., Willcox, D. C., & Suzuki, M. (2004). *The Okinawa diet plan*. Three Rivers Press.
61. Lopez-Huertas, E., & Fonolla, J. (2017). Hydroxytyrosol supplementation increases vitamin C levels *in vivo*: A human volunteer trial. *Redox Biology*, *11*, 384–389. https://doi.org/10.1016/j.redox.2016.12.014

CHAPTER 9: TO EVERYTHING THERE IS A SEASON

1. Li, H., Li, S., Yang, H., Zhang, Y., Zhang, S., Ma, Y., Hou, Y., Zhang, X., Niu, K., Borné, Y., & Wang, Y. (2022). Association of ultraprocessed food consumption with risk of dementia. A prospective cohort study. *Neurology*, *99*(10), e1056–e1066. https://doi.org/10.1212/WNL.0000000000200871
2. Goodpaster, B. H., & Sparks, L. M. (2017). Metabolic flexibility in health and disease. *Cell Metabolism*, *25*(5), 1027–1036. https://doi.org/10.1016/j.cmet.2017.04.015
3. Solanski, S., Sanchez, C., Ponnusamy, V., Kota, V., Bell, H. N., Cho, C.-S, Kowalsky, A. H., Green, M., Lee, J. H., & Shah, Y. M. (2023). Dysregulated amino acid sensing drives colorectal cancer growth and metabolic reprogramming leading to chemoresistance. *Gastroenterology*, *164*(3), 376–391.E13.
4. Kitada, M., Ogura, Y., Monno, I., & Koya, D. (2019). The impact of dietary protein intake on longevity and metabolic health, *EBioMedicine*, *43*, 632–640. https://doi.org/10.1016/j.ebiom.2019.04.005
5. Smits, S. A., Leach, J., Sonnenburg, E. D., Gonzalez, C. G., Lichtman, J. S., Reid, G., Knight, R., Manjurano, A., Changalucha, J., Elias, J. E., Dominguez-Bello, M. G., & Sonnenburg, J. L. (2017). Seasonal cycling in the gut microbiome of the Hadza hunter-gatherers of Tanzania. *Science*, *357*(6353), 802–806. https://doi.org/10.1126/science.aan4834
6. Zhang, Y., Zhang, J., & Wang, S. (2021). The role of rapamycin in healthspan extension via the delay of organ aging. *Ageing Research Reviews*, *70*, article 101376. https://doi.org/10.1016/j.arr.2021.101376
7. Harrison, D. E., Strong, R., Sharp, Z. D., Nelson, J. F., Astle, C. M., Flurkey, K., Nadon, N. L., Wilkinson, J. E., Frenkel, K., Carter, C. S., Pahor, M., Javors, M. A., Fernandez, E., & Miller, R. A. (2009). Rapamycin fed late in life extends lifespan in genetically heterogeneous mice. *Nature*, *460*(7253), 392–395.
8. Johnson, S. C., Yanos, M. E., Kayser, E. B., Quintana, A., Sangesland, M., Castanza, A., Uhde, L., Hui, J., Wall, V. Z., Gagnidze, A., Oh, K., Wasko,

B. M., Ramos, F. J., Palmiter, R. D., Rabinovitch, P. S., Morgan, P. G., Sedensky, M. M., & Kaeberlein, M. (2013). mTOR inhibition alleviates mitochondrial disease in a mouse model of Leigh syndrome. *Science*, *342*(6165), 1524–1528. https://doi.org/10.1126/science.1244360

9. Bitto, A., Ito, T. K., Pineda, V. V., LeTexier, N. J., Huang, H. Z., Sutlief, E., Tung, H., Vizzini, N., Chen, B., Smith, K., Meza, D., Yajima, M., Beyer, R. P., Kerr, K. F., Davis, D. J., Gillespie, C. H., Snyder, J. M., Treuting, P. M., & Kaeberlein, M. (2016). Transient rapamycin treatment can increase lifespan and healthspan in middle-aged mice. *Elife*, *5*, e16351. https://doi.org/10.7554/eLife.16351
10. Xu, L., Zhang, C., He, D., Jiang, N,. Bai, Y., & Xin, Y. (2020). Rapamycin and MCC950 modified gut microbiota in experimental autoimmune encephalomyelitis mouse by brain gut axis. *Life Sciences*, *253*, article 117747. https://doi.org/10.1016/j.lfs.2020.117747
11. Ke, H., Li, F., Deng, W., Li, Z., Wang, S., Lv, P., & Chen, Y. (2021). Metformin exerts anti-inflammatory and mucus barrier protective effects by enriching *Akkermansia muciniphila* in mice with ulcerative colitis. *Frontiers in Pharmacology*, *12*, article 726707. https://doi.org/10.3389/fphar.2021.726707
12. Aliper, A., Jellen, L., Cortese, F., Artemov, A., Karpinsky-Semper, D., Moskalev, A., Swick, A. G., & Zhavoronkov, A. (2017). Towards natural mimetics of metformin and rapamycin. *Aging*, *9*(11), 2245–2268. https://doi.org/10.18632/aging.101319
13. Dey, A., Chatterjee, S. S., & Kumar, V. (2018). Triethylene glycol–like effects of *Ashwagandha* (*Withania somnifera* (L.) Dunal) root extract devoid of withanolides in stressed mice. *Ayu*, *39*(4), 230–238. https://doi.org/10.4103/ayu.AYU_219_16
14. Lee, J. H., Budanov, A. V., & Karin, M. (2013). Sestrins orchestrate cellular metabolism to attenuate aging. *Cell Metabolism*, *18*(3), 792–801. https://doi.org/10.1016/j.cmet.2013.08.018
15. Lanna, A., Gomes, D.C.O., Muller-Durovic, B., McDonnell, T., Escors, D., Gilroy, D. W., Lee, J. H., Karin, M., & Akbar, A. N. (2017). A sestrin-dependent Erk-Jnk-p38 MAPK activation complex inhibits immunity during aging. *Nature Immunology*, *18*(3), 354–363.
16. Hu, H.-J., Shi, Z.-Y., Lin, X.-L., Chen, S.-M., Wang, Q.-Y., & Tang, S.-Y. (2015). Upregulation of Sestrin2 expression protects against macrophage apoptosis induced by oxidized low-density lipoprotein. *DNA and Cell Biology*, *34*(4), 296–302.
17. Ho, A., Cho, C.-S., Namkoong, S., Cho, U.-S., & Lee, J. H. (2016). Biochemical basis of Sestrin physiological activities. *Trends in Biochemical Sciences*, *41*(7), 621–632. https://doi.org/10.1016/j.tibs.2016.04.005
18. Budanov, A. V., Shoshani, T., Faerman, A., Zelin, E., Kamer, I., Kalinski, H., Grodin, S., Fishman, A., Chajut, A., Einat, P., Skaliter, R., Gudkov, A. V., Chumakov, P. M., & Feinstein, E. (2002). Identification of a novel stress-responsive gene Hi95 involved in regulation of cell viability. *Oncogene*, *21*(39), 6017–6031.
19. Peng, M., Yin, N., & Li, M. O. (2014). Sestrins function as guanine nucleotide

dissociation inhibitors for Rag GTPases to control mTORC1 signaling. *Cell, 159*(1),122–133. https://doi.org/10.1016/j.cell.2014.08.038
20. Green, C. L., & Lamming, D. W. (2019). Regulation of metabolic health by essential dietary amino acids. *Mechanisms of Ageing and Development*, *177*, 186–200.
21. Lee, J. H., Budanov, A. V., & Karin, M. (2013). Sestrins orchestrate cellular metabolism to attenuate aging. *Cell Metabolism*, *18*(3), 792–801. https://doi.org/10.1016/j.cmet.2013.08.018
22. Tao, R., Xiong, X., Liangpunsakul, S., & Dong, X. C. (2015). Sestrin 3 protein enhances hepatic insulin sensitivity by direct activation of the mTORC2-Akt signaling. *Diabetes*, *64*(4), 1211–1223.
23. Hallajzadeh, J., Eslami, R. D., & Tanomand, A. (2021). Effect of *Lactobacillus delbrueckii* subsp. *lactis* PTCC1057 on serum glucose, fetuin-A, and sestrin 3 levels in streptozotocin-induced diabetic mice. *Probiotics and Antimicrobial Proteins*, *13*(2), 383–389. https://doi.org/10.1007/s12602-020-09693-0
24. Kim, G. T., Lee, S. H., & Kim, Y. M. (2013). Quercetin regulates sestrin 2-AMPK-mTOR signaling pathway and induces apoptosis via increased intracellular ROS in HCT116 colon cancer cells. *Journal of Cancer Prevention*, *18*(3), 264–270. https://doi.org/10.15430/jcp.2013.18.3.264
25. Jin, S. H., Yang, J. H., Shin, B. Y., Seo, K., Shin, S. M., Cho, I. J., & Ki, S. H. (2013). Resveratrol inhibits LXRα-dependent hepatic lipogenesis through novel antioxidant *Sestrin2* gene induction. *Toxicology and Applied Pharmacology*, *271*(1), 95–105. https://doi.org/10.1016/j.taap.2013.04.023
26. Lu, Y.-X., Regan, J. C., Eßer, J., Drews, L. F., Weinseis, T., Stinn, J., Hahn, O., Miller, R. A., Grönke, S., & Partridge, L. (2021). A TORC1-histone axis regulates chromatin organisation and non-canonical induction of autophagy to ameliorate ageing. *eLife*, *10*, e62233. https://doi.org/10.7554/eLife.62233
27. Steliou, K., Boosalis, M. S., Perrine, S. P., Sangerman, J., & Faller, D. V. (2012). Butyrate histone deacetylase inhibitors. *BioResearch Open Access*, *1*(4), 192–198. https://doi.org/10.1089/biores.2012.0223
28. Blouin, J.-M., Penot, G., Collinet, M., Nacfer, M., Forest, C., Laurent-Puig, P., Coumoul, X., Barouki, R., Benelli, C., & Bortoli, S. (2011). Butyrate elicits a metabolic switch in human colon cancer cells by targeting the pyruvate dehydrogenase complex. *International Journal of Cancer*, *128*(11), 2591–2601.
29. Davie, J. R. (2003). Inhibition of histone deacetylase activity by butyrate. *The Journal of Nutrition*, *133*(7)(Suppl.), S2485S–S2493S. https://doi.org/10.1093/jn/133.7.2485S

CHAPTER 10: THE *GUT CHECK* EATING CYCLE

1. Li, B., Li, L., Li, M., Lam, S. M., Wang, G., Wu, Y., Zhang, H., Niu, C., Zhang, X., Liu, X., Hambly, C., Jin, W., Shui, G., & Speakman, J. R. (2019). Microbiota depletion impairs thermogenesis of brown adipose tissue and browning of white adipose tissue. *Cell Reports*, *26*(10), 2720–2737.e5. https://doi.org/10.1016/j.celrep.2019.02.015

2. Speakman, J. R., Talbot, D. A., Selman, C., Snart, S., McLaren, J. S., Redman, P., Krol, E., Jackson, D. M., Johnson, M. S., & Brand, M. D. (2004). Uncoupled and surviving: Individual mice with high metabolism have greater mitochondrial uncoupling and live longer. *Aging Cell*, *3*(3), 87–95. https://doi.org/10.1111/j.1474-9728.2004.00097.x
3. Peters, A., Krumbholz, P., Jäger, E., Heintz-Buschart, A., Çakir, M. V., Rothemund, S., Gaudl, A., Ceglarek, U., Schöneberg, T., & Stäubert, C. (2019). Metabolites of lactic acid bacteria present in fermented foods are highly potent agonists of human hydroxycarboxylic acid receptor 3. *PLOS Genetics*, *15*(5), e1008145. https://doi.org/10.1371/journal.pgen.1008145. Erratum (2019) in: *PLOS Genetics*, *15*(7), e1008283.
4. Taylor, B. C., Lejzerowicz, F., Poirel, M., Shaffer, J. P., Jiang, L., Aksenov, A., Litwin, N., Humphrey, G., Martino, C., Miller-Montgomery, S., Dorrestein, P. C., Veiga, P., Song, S. J., McDonald, D., Derrien, M., & Knight, R. (2020). Consumption of fermented foods is associated with systematic differences in the gut microbiome and metabolome. *mSystems*, *5*(2), e00901-19.

CHAPTER 11: *THE PLANT PARADOX 2.0*

1. Mu, G., Zhang, Z., Wang, J., Jiang, S., Wang, H., Xu, Y., Li, X., Chi, L., Li, Y., Tuo, Y., & Zhu, X. (2021). Antigenicity and safety evaluation of *Lactiplantibacillus plantarum* 7-2 screened to reduce α-casein antigen. *Foods*, *11*(1), 88. https://doi.org/10.3390/foods11010088
2. Bashir, S., Fezeu, L. K., Leviatan Ben-Arye, S., Yehuda, S., Reuven, E. M., Szabo de Edelenyi, F., Fellah-Hebia, I., Le Tourneau, T., Imbert-Marcille, B. M., Drouet, E. B., Touvier, M., Roussel, J. C., Yu, H., Chen, X., Hercberg, S., Cozzi, E., Soulillou, J. P., Galan, P., & Padler-Karavani, V. (2020). Association between Neu5Gc carbohydrate and serum antibodies against it provides the molecular link to cancer: French NutriNet-Santé study. *BMC Medicine*, *18*(1), 262. https://doi.org/10.1186/s12916-020-01721-8
3. Bashir, S., Fezeu, L. K., Leviatan Ben-Arye, S., Yehuda, S., Reuven, E. M., Szabo de Edelenyi, F., Fellah-Hebia, I., Le Tourneau, T., Imbert-Marcille, B. M., Drouet, E. B., Touvier, M., Roussel, J. C., Yu, H., Chen, X., Hercberg, S., Cozzi, E., Soulillou, J. P., Galan, P., & Padler-Karavani, V. (2020). Association between Neu5Gc carbohydrate and serum antibodies against it provides the molecular link to cancer: French NutriNet-Santé study. *BMC Medicine*, *18*(1), 262. https://doi.org/10.1186/s12916-020-01721-8
4. Davies, L.R.L., & Varki, A. (2015). Why is N-glycolylneuraminic acid rare in the vertebrate brain? *Topics in Current Chemistry*, *366*, 31–54. https://doi.org/10.1007/128_2013_419
5. Boligan, K. F., Oechtering, J., Keller, C. W., Peschke, B., Rieben, R., Bovin, N., Kappos, L., Cummings, R. D., Kuhle, J., von Gunten, S., & Lünemann, J. D. (2020). Xenogeneic Neu5Gc and self-glycan Neu5Ac epitopes are potential immune targets in MS. *Neurology: Neuroimmunology & Neuroinflammation*, 7(2), e676. https://doi.org/10.1212/NXI.0000000000000676
6. Naito-Matsui, Y., Davies, L. R., Takematsu, H., Chou, H. H., Tangvoranun-

takul, P., Carlin, A. F., Verhagen, A., Heyser, C. J., Yoo, S. W., Choudhury, B., Paton, J. C., Paton, A. W., Varki, N. M., Schnaar, R. L., & Varki, A. (2017). Physiological exploration of the long term evolutionary selection against expression of *N*-glycolylneuraminic acid in the brain. *Journal of Biological Chemistry*, *292*(7), 2557–2570. https://doi.org/10.1074/jbc.M116.768531

7. Le Berre, L., Salama, A., Evanno, G., Rousse, J., Nicot, A., Semana, G., Laplaud, D. A., Imbert, B.-M., Drouet, E., & Soulillou, J.-P. (2015). Increased IGM and IGG anti-NEU5GC antibodies in infectious mononucleosis (IMN): Link with multiple sclerosis (MS)? *Xenotransplantation*, *22*, S86.
8. Banda, K., Gregg, C. J., Chow, R., Varki, N. M., & Varki, A. (2012). Metabolism of vertebrate amino sugars with N-glycolyl groups: Mechanisms underlying gastrointestinal incorporation of the non-human sialic acid xeno-autoantigen N-glycolylneuraminic acid. *Journal of Biological Chemistry*, *287*(34), 28852–28864. https://doi.org/10.1074/jbc.M112.364182
9. Kawanishi, K., Coker, J. K., Grunddal, K. V., Dhar, C., Hsiao, J., Zengler, K., Varki, N., Varki, A., & Gordts, P.L.S.M. (2021). Dietary Neu5Ac intervention protects against atherosclerosis associated with human-like Neu5Gc loss—brief report. *Arteriosclerosis, Thrombosis, and Vascular Biology*, *41*(11), 2730–2739. https://doi.org/10.1161/ATVBAHA.120.315280
10. Kawanishi, K., Coker, J. K., Grunddal, K. V., Dhar, C., Hsiao, J., Zengler, K., Varki, N., Varki, A., & Gordts, P.L.S.M. (2021). Dietary Neu5Ac intervention protects against atherosclerosis associated with human-like Neu5Gc loss—brief report. *Arteriosclerosis, Thrombosis, and Vascular Biology*, *41*(11), 2730–2739. https://doi.org/10.1161/ATVBAHA.120.315280
11. Samraj, A. N., Läubli, H., Varki, N., & Varki, A. (2014). Involvement of a non-human sialic acid in human cancer. *Frontiers in Oncology*, *4*, 33. https://doi.org/10.3389/fonc.2014.00033
12. Lin, X., Yao, H., Guo, J., Huang, Y., Wang, W., Yin, B., Li, X., Wang, T., Li, C., Xu, X., Zhou, G., Voglmeir, J., & Liu, L. (2022). Protein glycosylation and gut microbiota utilization can limit the in vitro and in vivo metabolic cellular incorporation of Neu5Gc. *Molecular Nutrition & Food Research*, *66*(5), article 2100615.
13. Wu, G., & Li, P. (2022). The "ideal protein" concept is not ideal in animal nutrition. *Experimental Biology and Medicine*, *247*(13), 1191–1201. https://doi.org/10.1177/15353702221082658
14. Hu, S., He, W., & Wu, G. (2022). Hydroxyproline in animal metabolism, nutrition, and cell signaling. *Amino Acids*, *54*(4), 513–528. https://doi.org/10.1007/s00726-021-03056-x
15. Eutamene, H., Beaufrand, C., Harkat, C., & Theodorou, V. (2022). Effect of two mucoprotectants, gelatin tannate and xyloglucan plus gelatin, on cholera toxin–induced water secretion in rats. *Gastrointestinal Disorders*, *4*(4), 324–332. https://doi.org/ 10.3390/gidisord4040030
16. Gundry, S. R. (2018). Remission/cure of autoimmune diseases by a lectin limited diet supplemented with probiotics, prebiotics, and polyphenols. *Circulation*, *137*(Suppl. 1), abstract AP238.
17. Labrada, M., Dorvignit, D., Hevia, G., Rodríguez-Zhurbenko, N., Hernández,

A. M., Vázquez, A. M., & Fernández, L. E. (2018). GM3(Neu5Gc) ganglioside: An evolution fixed neoantigen for cancer immunotherapy. *Seminars in Oncology*, *45*(1–2), 41–51. https://doi.org/10.1053/j.seminoncol.2018.04.003

CHAPTER 12: THE *GUT CHECK* FOOD PLAN

1. Bernardi, S., Del Bo', C., Marino, M., Gargari, G., Cherubini, A., Andrés-Lacueva, C., Hidalgo-Liberona, N., Peron, G., González-Dominguez, R., Kroon, P., Kirkup, B., Porrini, M., Guglielmetti, S., & Riso, P. (2020). Polyphenols and intestinal permeability: Rationale and future perspectives. *Journal of Agricultural and Food Chemistry*, *68*(7), 1816–1829. https://doi.org/10.1021/acs.jafc.9b02283
2. Pérez-Jiménez, J., Neveu, V., Vos, F., & Scalbert, A. (2010). Identification of the 100 richest dietary sources of polyphenols: An application of the phenol-explorer database. *European Journal of Clinical Nutrition*, *64*(Suppl.), S112–S120.
3. Sharma, R., Diwan, B., Singh, B. P., Kulshrestha, S. (2022). Probiotic fermentation of polyphenols: Potential sources of novel functional foods. *Food Production, Processing and Nutrition*, *4*, article 21. https://doi.org/10.1186/s43014-022-00101-4
4. Hahn, J., Cook, N. R., Alexander, E. K., Friedman, S., Walter, J., Bubes, V., Kotler, G., Lee, I. M., Manson, J. E., & Costenbader, K. H. (2022). Vitamin D and marine omega 3 fatty acid supplementation and incident autoimmune disease: VITAL randomized controlled trial. *The British Medical Journal*, *376*, e066452. https://doi.org/10.1136/bmj-2021-066452
5. Ghahremani, M., Smith, E. E., Chen, H.-Y., Creese, B., Goodarzi, Z., & Ismail, Z. (2023). Vitamin D supplementation and incident dementia: Effects of sex, *APOE*, and baseline cognitive status. *Alzheimer's & Dementia*, *15*(1), e12404. https://doi.org/10.1002/dad2.12404
6. Garland, C. F., French, C. B., Baggerly, L. L., & Heaney, R. P. (2011). Vitamin D supplement doses and serum 25-hydroxyvitamin D in the range associated with cancer prevention. *Anticancer Research*, *31*(2), 607–611.
7. Tsaban, G., Shalev, A., Katz, A., Meir, A. Y., Rinott, E., Zelicha, H., Kaplan, A., Wolak, A., Bluher, M., Stampfer, M. J., & Shai, I. (2023). Effect of lifestyle modification and green Mediterranean diet on proximal aortic stiffness. *Journal of the American College of Cardiology*, *81*(16), 1659–1661. https://doi.org/10.1016/j.jacc.2023.02.032
8. Bradman, A., Quirós-Alcalá, L., Castorina, R., Schall, R. A., Camacho, J., Holland, N. T., Barr, D. B., & Eskenazi, B. (2015). Effect of organic diet intervention on pesticide exposures in young children living in low-income urban and agricultural communities. *Environmental Health Perspectives*, *10*, 1086–1093. https://doi.org/10.1289/ehp.1408660
9. Li, Y., Lai, W., Zheng, C., Babu, J. R., Xue, C., Ai, Q., & Huggins, K. W. (2022). Neuroprotective effect of stearidonic acid on amyloid β–induced neurotoxicity in rat hippocampal cells. *Antioxidants*, *11*(12), 2357. https://doi.org/10.3390/antiox11122357

10. Kawamura, A., Nemoto, K., & Sugita, M. (2023). Effect of 8-week intake of the n-3 fatty acid—rich perilla oil on the gut function and as a fuel source for female athletes: A randomised trial. *British Journal of Nutrition*, *129*(6), 981–991. https://doi.org/10.1017/S0007114522001805
11. Kawashima, H. (2019). Intake of arachidonic acid–containing lipids in adult humans: Dietary surveys and clinical trials. *Lipids in Health and Disease*, *18*, 101. https://lipidworld.biomedcentral.com/articles/10.1186/s12944-019-1039-y
12. Fasano, A. (2020). All disease begins in the (leaky) gut: Role of xonulin-mediated gut permeability in the pathogenesis of some chronic inflammatory diseases. *F1000 Research*, *9*. https://www.ncbi.nlm.nih.gov/pmc/articles/PMC6996528/
13. Peh, M. T., Anwar, A. B., Ng, D.S.W., Atan, M.S.B.M., Kumar, S. D., & Moore, P. K. (2014). Effect of feeding a high fat diet on hydrogen sulfide (H_2S) metabolism in the mouse. *Nitric Oxide*, *41*, 138–145. https://europepmc.org/article/med/24637018
14. Pinget, G., Tan, J., Janec, B., Kaakoush, N. O., Angelatos, A. S., O'Sullivan, J., Koay, Y. C., Sierro, F., Davis, J., Divakarla, S. K., Khanal, D., Moore, R. J., Stanley, D., Chrzaowski, W., & Macia, L. (2019). Impact of the food additive titanium dioxide (E171) on gut microbiota–host interaction. *Frontiers in Nutrition*, 6, article 57. https://www.frontiersin.org/articles/10.3389/fnut.2019.00057/full
15. Fauste, E., Donis, C., Panadero, M. I., Otero, P., & Bocos, C. (2021, June 1). Fructose Consumption Hampers Gasotransmitter production. *Academia Letters*, article 1380. https://doi.org/10.20935/AL1380.3
16. Crescenzo, R., Mazzoli, A., Di Luccia, B., Bianco, F., Cancelliere, R., Cigliano, L., Liverin, G., Baccigalupi, L., & Iossa, S. (2017). Dietary fructose causes defective insulin signalling and ceramide accumulation in the liver that can be reversed by gut microbiota modulation. *Food & Nutrition Research*, *61*(1), article 1331657. https://doi.org/10.1080/16546628.2017.1331657
17. Olson, E., Suh, J. H., Schwarz, J.-M., Noworolski, S. M., Jones, G. M., Barber, J. R., Erkin-Cakmak, A., Mulligan, K., Lustig, R. H., & Mietus-Snyder, M. (2022). Effects of isocaloric fructose restriction on ceramide levels in children with obesity and cardiometabolic risk: Relation to hepatic de novo lipogenesis and insulin sensitivity. *Nutrients*, *14*(7), 1432. https://doi.org/10.3390/nu14071432
18. Han, Y., Kwon, E.-Y., & Choi, M.-S. (2020). Anti-diabetic effects of allulose in diet-induced obese mice via regulation of mRNA expression and alteration of the microbiome composition. *Nutrients*, *12*(7), 2113. https://doi.org/10.3390/nu12072113
19. Han, Y., Park, H., Choi, B.-R., Ji, Y., Kwon, E.-Y., & Choi, M.-S. (2020). Alteration of microbiome profile by D-allulose in amelioration of high-fat-diet-induced obesity in mice. *Nutrients*, *12*(2), 352. https://doi.org/10.3390/nu12020352
20. Yao, C. K., Muir, J. G., and Gibson, P. R. (2016). Review article: Insights into colonic protein fermentation, its modulation and potential health implications. *Alimentary Pharmacology & Therapeutics*, *43*, 181–196. https://onlinelibrary.wiley.com/doi/pdf/10.1111/apt.13456

21. David, L. A., Maurice, C. F., Carmody, R. N., Gootenberg, D. B., Button, J. E., Wolfe, B. E., Ling, A. V., Devlin, A. S., Varma, Y., Fischbach, M. A., Biddinger, S. B., Dutton, R. J., & Turnbaugh, P. J. (2014). Diet rapidly and reproducibly alters the human gut microbiome. *Nature*, *505*(7484), 559–563. https://pubmed.ncbi.nlm.nih.gov/24336217/
22. Yao, C. K., Muir, J. G., and Gibson, P. R. (2016). Review article: Insights into colonic protein fermentation, its modulation and potential health implications. *Alimentary Pharmacology & Therapeutics*, *43*, 181–196. https://onlinelibrary.wiley.com/doi/pdf/10.1111/apt.13456

INDEX

ABOUT THE AUTHOR

STEVEN R. GUNDRY, MD, is the director of the International Heart and Lung Institute in Palm Springs, California, and the founder and director of the Centers for Restorative Medicine in Palm Springs, Santa Barbara, and Beverly Hills. He is the cofounder of GundryHealth.com, his telemedicine portal for the treatment of autoimmune diseases, IBS, and leaky gut. After a distinguished surgical career as a professor of surgery and pediatrics and chairman of cardiothoracic surgery at Loma Linda University School of Medicine, Dr. Gundry changed his focus to curing modern diseases via dietary changes and supplementation. He is the author of multiple *New York Times* bestsellers, is the host of the top-rated *Dr. Gundry Podcast*, and is the cofounder of GundryMD.com, his supplement, skin care, and food company. He has published more than three hundred articles in peer-reviewed journals on using diet and supplements to eliminate heart disease, diabetes, autoimmune disease, intestinal permeability, and multiple other conditions. Dr. Gundry lives with his wife, Penny, and their four dogs (including two new rescues) in Palm Springs and Montecito, California.